SPECIAL PAPERS IN PALAEONTOLOGY NO. 92

TRILOBITES FROM SILURIAN REEFS IN NORTH GREENLAND

BY

HELEN E. HUGHES *and* ALAN T. THOMAS

with 36 figures and 2 tables

THE PALAEONTOLOGICAL ASSOCIATION
LONDON

November 2014

CONTENTS

[Special Papers in Palaeontology, 92, 2014, pp. 5–102]

Abstract: Varied and well preserved trilobite faunas are described from Telychian (Llandovery, Silurian) reefs in North Greenland. The faunas, collected between Kronprins Christian Land in the east and Wulff Land some 600 km farther west, comprise 26 named species (23 new) and 39 under open nomenclature. These are assigned to 29 genera (four new). Members of the Illaenidae (one new species), Scutelluidae (nine new species, two new genera), Proetidae (seven new species, two new genera), Tropidocoryphidae (one new species), Aulacopleuridae, Scharyiidae, Cheiruridae (two new species), Encrinuridae (two new species), Calymenidae, Phacopidae, Lichidae (one new species), Odontopleuridae and Harpetidae are represented. The large number of new taxa reflects the previous lack of sampling of Silurian reef trilobite biotas. In the North Greenland reefs, trilobites occur predominantly in a cement-rich microbial lithofacies deposited between storm and fair-weather wave base, the fossils being concentrated in cavities and depressions on the reef surface. Several associations are recognized, which can be encompassed within the previously defined and long-ranging Illaenid–Cheirurid 'Community': the faunas have particularly close affinities with other Telychian Laurentian faunas described from similar environments. Species are not evenly distributed across the reef belt, however, and this smaller-scale variation is likely age-related, possibly reflecting the gradual foundering of the carbonate platform from east to west.

Key words: Greenland, reef, Silurian, Telychian, Trilobita, carbonates.

THE faunas described here from the Telychian reefs of North Greenland (Fig. 1) were collected during the North Greenland reconnaissance geological mapping programme of the Geological Survey of Greenland (at the time GGU, now GEUS) in the 1970s and 1980s. Palaeontological samples were taken when and where practicable, rather than as part of a systematic sampling programme. The reef tract exposed across North Greenland represents the oldest large-scale reef development in the Silurian and includes the most extensive of the late Llandovery reef complexes (Copper and Brunton 1991). The reefs have yielded thousands of predominantly disarticulated calcitic trilobite sclerites which are dominated by scutelluids (Hughes and Thomas 2011). These sclerites, like the reefs themselves, are remarkably well preserved, providing the most significant trilobite collections so far from any early Silurian reefs. The reef faunas from the Great Lakes region of the USA, which are geographically nearest, are predominantly both younger (Wenlock) and pervasively dolomitized (Lowenstam 1950, 1957; Copper and Brunton 1991).

The distribution of Silurian trilobite faunas is strongly related to lithofacies (Thomas 1979; Mikulic 1981, 1999; Männil 1982a, b; Chlupáč 1987; Thomas and Lane 1999). Trilobites in the North Greenland reefs are mostly associated with a cement-rich microbial lithofacies and were deposited between storm and fair-weather wave base, in cavities and depressions within the reef surface (Hughes and Thomas 2011). The faunas found belong to the Illaenid–Cheirurid Community (Fortey 1975), an umbrella term used to denote recurrent groups of trilobites thought to have lived together. The term community, in its biological sense, cannot be applied strictly to fossil assemblages as it refers to populations of organisms actually alive at a particular time; but is used here in accordance with palaeontological convention to denote the broader, sometimes long-ranging, associations of trilobites recognized from similar palaeoenvironments.

The large number of new species described here reflects the lack of sampling of Silurian reef environments, both in North Greenland and elsewhere. The financial and logistical constraints of conducting fieldwork in such a remote area have restricted collecting along the North Greenland reef belt. Other Telychian reefs are also most prevalent in inaccessible regions: other areas of the Arctic, for example (Copper and Brunton 1991), and the Gorny Altai, SW Siberia (Sennikov et al. 2008). Previous works concentrating on trilobite taxa from strata of similar age and lithofacies are few (Lane 1972, 1979, 1984; Norford 1973, 1981; Lane and Owens 1982; Adrain et al. 1995; Chatterton and Ludvigsen 2004). Other significant studies of Telychian trilobite faunas predominantly concern those from

HELEN E. HUGHES

School of Geography, Earth and Environmental Sciences, Plymouth University, Drake Circus, Plymouth, PL4 8AA, UK; e-mail: helen.hughes@plymouth.ac.uk

ALAN T. THOMAS

School of Geography, Earth and Environmental Sciences, University of Birmingham, Edgbaston, Birmingham, B15 2TT, UK; e-mail: a.t.thomas@bham.ac.uk

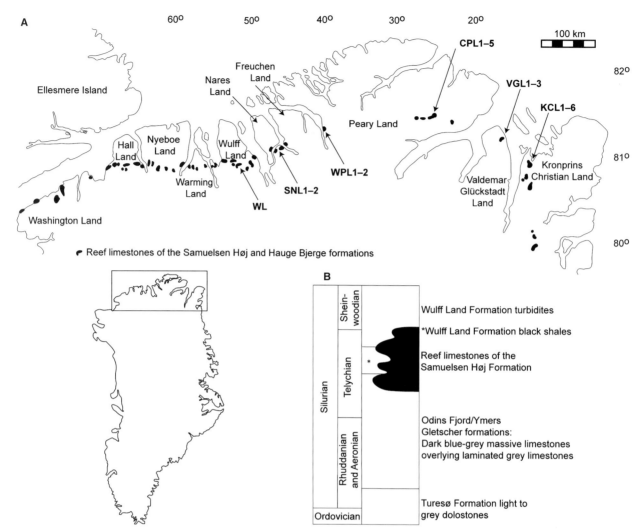

FIG. 1. A, map of North Greenland showing distribution of Silurian reefs (in black), adapted from Sønderholm and Harland (1989). Sample localities are indicated. B, summary stratigraphy of Central Peary Land showing the relationship of Telychian reefs to other stratigraphical units. Trilobites occur in the Samuelsen Høj and Odins Fjord formations and their time equivalents. For details of stratigraphical nomenclature, see Hurst (1984), Sønderholm *et al.* (1987) and Peel and Sønderholm (1991). After Hughes and Thomas (2011).

deeper-water lithofacies (Chatterton and Perry 1983, 1984; Ludvigsen and Tripp 1990; Norford 1994).

PREVIOUS WORK

The first palaeontological work to consider the Silurian trilobite faunas of North Greenland was that of Poulsen (1934), who studied the fossils collected by Dr Lauge Koch during the Second Thule Expedition (1916–1918) and the Danish Jubilee Expedition (1920–1923). The material available to Poulsen was from the late Rhuddanian – Aeronian Cape Schuchert Formation, Washington Land, and contained representatives of the Harpetidae, Aulacopleuridae, Proetidae, Odontopleuridae, Scutellui-

dae, Cheiruridae and Encrinuridae. Most of the taxa are not congeneric with those described here, and this probably reflects both the older age of the formation and its lithofacies, which Hurst (1980) documented as being dominated by bituminous cherty lime mudstones. However, Poulsen's fauna compares very closely with that described by Ludvigsen and Tripp (1990) from the relatively deep-water, dark grey argillaceous lime mudstone lithofacies of the early Silurian of the Road River Group at Prongs Creek, Yukon Peninsula. Teichert (1937), in a study of mainly Canadian Arctic collections resulting from the Fifth Thule Expedition (1921–1924), also described the species now known as *Opoa magnifica*, from the Telychian Offley Island Formation, St George Fjord, North Greenland.

Over 30 years then passed until material collected from Kronprins Christian Land (north-east Greenland) during the Lauge Koch expeditions was studied by Lane (1972). The material was collected from repetitive sequences of dolostones, limestones and shales then thought to belong to the Centrum Limestone and Dolomite, and the Drommebjerg Limestone. These stratigraphical terms are no longer in use, and were employed at a time before the effects of multiple thrusting were recognized in the area (Smith and Rasmussen 2008). The material probably came from the Samuelsen Høj Formation. New species of scutelluids, harpetids and cheirurids were described, as well as unnamed forms of aulacopleurids, calymenids, 'goldillaenids', illaenids, lichids, odontopleurids and proetids. Lane discussed the similarities between his fauna and others of early Ordovician to mid-Devonian age, particularly those of the Niagaran ('middle' Silurian) reefs of the Great Lakes, New York, the Lower Devonian Koněprusy Limestone of the Czech Republic and the Middle Devonian limestones from South Devon, England. He noted that trilobite faunas might be expected to show a greater overall similarity within pure limestone rocks of different ages, than in rocks of different lithofacies of the same age. Ten genera and three species described by Lane (1972) are represented in the faunas documented here.

Expeditions in the 1960s to 1980s (regional geological mapping project of GGU) resulted in the collection of numerous fossiliferous samples. Material from the Telychian Odins Fjord Formation in Peary Land was studied by Lane (1988) and from a more northerly locality by Thomas (1994). No taxa from the two localities are congeneric, but when compared with a sample collected from Wulff Land studied here from the time equivalent of the Odins Fjord Formation (the Djævlekløften Formation), broader similarities are evident. The Odins Fjord and Djævlekløften formations are the only Silurian formations from North Greenland yielding members of the Phacopidae and Pterygometopidae, reflecting the difference between the level-bottom-dwelling faunas of those formations and the reef faunas of the Samuelsen Høj Formation.

Trilobite faunas from western North Greenland have been studied by: Norford (1973) who described species of *Scotoharpes* from the Offley Island Formation at Kap Tyson and Kap Schuchert; Lane (1979) from the Cape Schuchert Formation in Washington Land; Lane and Owens (1982) from blocks derived from the Telychian Pentamerus Bjerg Formation in Washington Land; Lane (1984) from the Telychian Offley Island and Hauge Bjerge formations in Hall Land and Nyeboe Land; and Lane and Siveter (1991) from the Telychian Offley Island Formation, Kap Tyson, Hall Land. Of these, the material of Lane and Owens (1982) and Lane (1984) compares most closely with that described here, with six and five

congeneric taxa, respectively. This is a reflection of their similar age and lithofacies. The fauna studied by Lane and Siveter (1991) is unusual as it is dominated by *Calymene*. It is found in the time equivalent of the Odins Fjord and Djævlekløften formations, further demonstrating the faunal differences between level-bottom and reef environments.

STRATIGRAPHY AND GEOLOGICAL SETTING

A thick east–west striking succession of Mesoproterozoic to Tertiary rocks is exposed in North Greenland (Henriksen *et al.* 2009). The Lower Palaeozoic strata represent the sedimentary fill of the Franklinian Basin, which extends some 2000 km from the Canadian Arctic Islands to eastern North Greenland (Peel and Sønderholm 1991; Henriksen *et al.* 2009) and formed part of the Laurentian passive margin of the Iapetus Ocean. It is now widely agreed that Laurentia occupied an equatorial position throughout the early Palaeozoic, largely between 30 degrees north and 30 degrees south (Scotese and McKerrow 1990; Harper *et al.* 1996; Torsvik *et al.* 1996; MacNiocaill *et al.* 1997; Cocks and Torsvik 2002; Fortey and Cocks 2003). This is indicated by both palaeomagnetic and zoogeographical data (Smith and Rasmussen 2008).

From the late early Cambrian to the early Silurian, carbonate platform sedimentation dominated the Franklinian Basin (Henriksen *et al.* 2009). Different lithostratigraphical names have been given to units of similar lithofacies and age occurring across North Greenland. The summary which follows uses the terminology of Hurst (1984) which is applied to strata between Kronprins Christian Land and Central Peary Land in eastern North Greenland. The lower Llandovery comprises variable grey and dark grey limestones and dolostones belonging to the Turesø, Ymers Gletscher and Odins Fjord formations, displaying varying degrees of silicification. These pass upwards into Telychian argillaceous limestones and black shales of the Wulff Land Formation. Reefs are associated with these strata (Fig. 1B). The reefs are thought to have been initiated due to relative sea level rise (Hurst 1980), by means of either eustatic or continental transgression, or shelf and basin subsidence related to synsedimentary faulting, or both.

The reefs form a roughly west–east trending belt (Fig. 1A). Dawes (1971, 1976) first reported the reefs to cover a distance of *c.* 300 km, from Washington Land in the west to the Victoria Fjord region in the east. However, reefs are now known to extend across the whole of North Greenland (850 km), from Washington Land in the west to Kronprins Christian Land in the east

(Sønderholm and Harland 1989). The most southerly located reef in this belt was recognized in 1994–1995, in the south of Kronprins Christian Land (Smith *et al.* 2004). Reefs are variable in size; Lane and Thomas (1979) recorded a maximum thickness of 300 m, which, due to the onlapping nature of the overlying graptolitic shales and surrounding turbidites, is interpreted as original relief. In Central Peary Land, Valdemar Glückstadt Land and Kronprins Christian Land, the reefs comprise the Samuelsen Høj Formation, and in Western Peary Land and South Nares Land, they comprise the Hauge Bjerge Formation. Thin sections were made from samples from all localities, and there are no significant differences in the reef lithofacies, other than a slightly greater amount of dolomite in those of the Hauge Bjerge Formation. These formations have been little studied with respect to their sedimentology and petrography. Mayr (1976) originally described two of the Peary Land reefs, followed by Christie and Peel (1977), Lane and Thomas (1979) and Mabillard (1980). The reef lithofacies was described by Lane and Thomas (1979) as white to light grey crinoidal and stromatoporoidal bioclastic limestone. Bedding was described as massive in the core rocks and highly variable in the outer parts, and depositional dips were found to radiate outwards from the reef cores. The dominant fossil groups recorded were stromatoporoids and crinoid fragments, with pockets of other fossils including brachiopods, rugose corals, tabulate corals, trilobites, gastropods, cephalopods, rostroconchs and bivalves. Hughes and Thomas (2011) found cement-rich microbial boundstones to constitute the predominant lithofacies, and these, together with subordinate crinoidal grainstones, to yield the trilobite faunas. Minor coral boundstones are also present.

Foundering of the platform from late Llandovery times onwards was due to loading by thrust sheets from the north, and the platform carbonate deposition was replaced by turbidite deposition (Higgins *et al.* 1991). This resulted in most of the reef developments terminating during the late Llandovery, with those in western North Greenland persisting into the early Ludlow. At their stratigraphically highest points, reefs of the Samuelsen Høj Formation are surrounded by sand-rich turbidite deposits (Dawes 1976).

All material was collected from the Samuelsen Høj/Hauge Bjerge formations, except for that from Wulff Land which was collected from contemporaneous strata between reefs (the Djævlekløften Formation, the time equivalent of the Odins Fjord Formation).

AGE OF THE FAUNAS

The accurate dating of the reef limestones is hampered by a lack of conodonts. However, in eastern North Greenland, conodonts have been collected from the Odins Fjord Formation which belong to the *Pterospathodus celloni* Biozone (Aldridge 1979), and Armstrong (1990) recorded *celloni* as the youngest fauna from the reefs. The *celloni* Biozone lies within the upper but not uppermost Telychian, and is equivalent to the lower *Oktavites spiralis* graptolite Biozone (Ogg *et al.* 2008). All material described here is considered to be Telychian in age.

MATERIAL AND LOCALITY INFORMATION

Material

In total, over 400 kg of bulk limestone collected from North Greenland between 1978 and 1985 was available for study. This was broken down mechanically to allow detailed preparation of the calcitic sclerites (see Hughes and Thomas 2011, for details of sample processing). Although the trilobites are predominantly preserved with cuticle, this adheres to the external mould and is therefore often unavoidably lost with the matrix during preparation with a pneumatic air pen.

Locality information

Bulk rock samples are distinguished by their 'GGU' numbers which relate directly to the collection localities. GGU samples from the same reef localities are grouped together with an abbreviated locality designation and a number to distinguish between reefs; for example, WPL1 denotes Western Peary Land, reef 1. Details of the GGU numbers concerned are given in Table 1, and collection localities are shown in Figure 1A.

Repositories

All type and figured specimen numbers are prefixed with 'MGUH' and are housed in the Geological Museum, Natural History Museum of Denmark, Copenhagen, in accordance with established practice. Further reef limestone samples from Western Peary Land (WPL1), Wulff Land and South Nares Land comprise the Peel Collection and are housed in the Geological Museum, Natural History Museum of Denmark, Copenhagen. Other specimens, from Western Peary Land (WPL2), central Peary Land, Kronprins Christian Land and Valdemar Glückstadt Land, are housed at the Lapworth Museum, University of Birmingham. Curatorial practice there is to allocate 'BU' numbers only to type and figured specimens, with 'BIRUG' numbers being used for other particularly well

TABLE 1. Details of GGU numbers studied.

Geographical location	Abbreviation	GGU numbers	Map reference	Collector	Year collected	Comments from field notebooks
Wulff Land	WL	298504–298506	GGU map sheet 81V3	JSP	1985	From the inter-reef Djævlekløften formation
South Nares Land	SNL1	298930	Not available	JSP	1985	East of Peel Camp 6 from sides of main mound
	SNL2	298924	Not available	JSP	1985	Downslope main valley of Peel Camp 6
Western Peary Land	WPL1	301314, 301318–9, 301323–6	Not available	JSP	1985	South-east side of JP Koch Fjord
	WPL2	198833–6, 198838, 198842–6	Not available	JMH	1978	JP Koch Fjord
Central Peary Land	CPL1	198115–6, 198226	27Y UM 7573	ATT, PDL	1978	
	CPL2	198135–198223	27Y UM 7572	ATT, PDL	1978	
	CPL2	274645	27Y UM 7572	PDL	1980	Main mound, slope of hill at southern extremity of south-east limb
	CPL3	198225	27Y UM 804 679	ATT	1978	
	CPL4	274654	Not available	PDL	1980	Loose material from 'Mab's mound' (large mound E of Samuelsen Høj)
	CPL5	274689	Not available	HAA, PDL	1980	*In situ* and loose material from 'sigmoid mound'
Valdemar Glükstadt Land	VGL1	275038	Not available	PDL, JSP	1980	*In situ* and loose material from near Lane/Peel Camp 14
	VGL2	275044	Not available	PDL	1980	*In situ* material from 10 km south of Lane/Peel Camp 14
	VGL3	275049	Not available	PDL, JSP	1980	*In situ* and loose material from 3 km west of Lane/Peel Camp 14
Kronprins Christian Land	KCL1	275015	Not available	HAA, JSP	1980	Reef rock near Peel/Armstrong Camp 11
	KCL2	225793	Not available	JSP	1979	
	KCL3	275021	Not available	TW	1980	Mound collection near Peel/Armstrong Camp 11
	KCL4	274996	Not available	PDL, JSP	1980	Vertical reef, 1 km south-east of Lane/Peel Camp 8
	KCL5	274774	Not available	PDL	1980	Large mound mass 1 km west of Lane/Peel Camp12
	KCL6	274788	Not available	PDL	1980	1 km north of Lane/Peel Camp 8

All available details of locality information have been included. Initials for collectors are as follows: JSP, Dr. J. S. Peel; JMH, Dr J. M. Hurst; ATT, Dr A. T. Thomas; PDL, Dr P. D. Lane; HAA, Dr H. A. Armstrong; TW, Dr T. Winsnes.

preserved specimens. As a matter of practicality therefore, specimen numbers are quoted here only for holotypes and figured paratypes. The number of unfigured paratypes for the species described can be determined from the data in Table 2, and the specimens themselves can be identified from the Lapworth Museum catalogue.

ALPHA DIVERSITY

Table 2 shows the distribution of the trilobite taxa described here (as numbers of individuals) across the North Greenland reef belt. A clear faunal distinction can be made between trilobite taxa occurring in Western

TABLE 2. Distribution of trilobite taxa across North Greenland. Numbers given are of trilobite individuals, calculated from the greater of the following groupings: (cephala + cranidia + cephalothoraces) and (pygidia + thoracopygidia).

Taxon	WEST					EAST
	Locality					
	WL	SNL	WPL	CPL	VGL	KCL
Stenopareia persica sp. nov.	0	0	0	43	1	0
Stenopareia cf. grandis	5	1	20	0	0	0
Stenopareia sp.	0	0	0	0	2	0
Ekwanoscutellum agmen	0	0	0	351	7	18
Jungosulcus emilyae	0	0	0	203	0	0
Jungosulcus anaphalantos	0	226	0	0	0	0
Jungosulcus ventricosus	0	0	0	0	25	0
Meroperix cf. ataphrus	0	0	0	0	1	0
Opoa limatula	0	0	0	23	0	3
Opoa magnifica	99	34	99	0	0	0
Opoa cf. magnifica	0	0	0	2	0	0
Opoa sp.	0	0	0	0	0	2
Scutelluid gen. indet. 1	0	0	0	0	0	2
Scutelluid gen. indet. 2	0	0	2	0	0	0
Cybantyx nebulosus	0	0	0	127	0	0
Cybantyx sp.	0	2	14	0	0	0
Laneia enalios	0	0	0	0	11	0
Ligiscus diana	0	0	0	3	65	3
Liolalax naresi	0	14	0	0	0	0
Liolalax? sp.	0	0	0	0	1	1
Proetus? confluens	0	0	0	12	0	0
Airophrys balia	0	0	0	0	1	0
Airophrys sp. 1	0	0	0	3	0	0
Cyphoproetus sp. 1	0	0	0	4	0	0
Cyphoproetus sp. 2	0	1	0	0	0	0
Owensus arktoperates	0	0	0	74	0	0
Astroproetus franklini	0	0	0	4	1	0
Astroproetus sp.	0	0	0	0	0	1
Thebanaspis sp.	1	0	1	0	0	0
Winiskia eruga	0	0	0	5	0	3
Winiskia leptomedia	0	0	0	0	0	3
Winiskia stickta	0	0	0	7	0	0
Winiskia sp.	0	0	0	5	0	0
Dalarnepeltis brevifrons	0	0	0	5	0	0
Tropidocoryphine? gen. et sp. indet.	0	0	0	2	0	0

TABLE 2. (Continued)

Taxon	WEST					EAST
	Locality					
	WL	SNL	WPL	CPL	VGL	KCL
Aulacopleurid cf. Songkania	0	0	0	0	1	0
Scharyia sp. 1	0	0	0	0	1	0
Scharyia sp. 2	0	0	0	9	0	0
Scharyia sp. 3	0	0	0	2	0	0
Cheirurus falcatus	0	0	0	52	1	2
Proromma? sp.	0	0	0	2	5	1
Radiurus pauli	58	15	14	0	0	0
Hyrokybe pharanx?	0	0	0	2	0	0
Hyrokybe? sp.	0	0	0	7	0	0
Acanthoparyphinae gen. et sp. indet. 1	0	0	0	1	0	0
Acanthoparyphinae gen. et sp. indet. 2	0	0	0	1	0	0
Distyrax bibullatus	0	0	0	0	0	1
Distyrax? sp.	0	0	0	1	0	0
Perryus mikulici	0	0	2	35	0	37
Perryus cf. P. palasso	2	1	0	0	0	0
Perryus sp. 1	0	0	0	2	0	0
Perryus sp. 2	0	0	0	0	0	2
Genus aff. Perryus	0	0	0	7	0	0
variolaris plexus gen. indet. 1	0	0	0	2	0	1
variolaris plexus gen. indet. 2	0	0	0	21	0	0
Calymene aff. iladon	4	0	0	0	0	0
Calymene sp.	0	0	0	6	0	0
Calymenina indet.	0	0	1	0	0	0
Acernaspis? sp.	1	0	0	0	0	0
Dicranogmus pearyi	0	0	0	13	0	0
Dicranogmus sp.	0	0	0	15	0	0
Trochurinae indet.	0	0	0	2	0	0
Ceratocephala sp. 1	0	0	0	8	0	1
Ceratocephala sp. 2	2	0	0	0	0	0
Scotoharpes loma	2	29	14	9	2	5

Peary Land and further west, and those occurring in Central Peary Land and further east. Only two species, *Scotoharpes loma* and *Perryus mikulici*, occur in reefs from both of these broad areas. The termination of reef development by the onset of deeper-water sedimentation has been shown to young from east to west across the North Greenland carbonate platform by limited graptolite evidence (Smith and Rasmussen 2008). This could mean that the faunas occurring in the west of North Greenland are slightly younger than those from the most easterly reefs.

PALAEOBIOGEOGRAPHY

The trilobite faunas from the reefs of North Greenland have close affinities with those of similar age described from other northern Laurentian shallow marine environments. Trilobites from the Telychian – ?lower Wenlock Attawapiskat Formation from northern Ontario and northern Manitoba (Norford 1981; Gass and Mikulic 1982; Westrop and Rudkin 1999) occur in pockets within reef lithofacies. Seven of the 15 genera described by Norford (1981) are represented in North Greenland. North Laurentian links are indicated also by a Telychian trilobite fauna described from Alaska by Adrain *et al.* (1995); the seven species identified from the white bioclastic limestone from the Taylor Mountains are all congeneric with those from North Greenland. Chatterton and Ludvigsen (2004) revised and described all known Llandovery trilobites from Anticosti Island, Québec, Canada. They identified six successive faunas. There are some generic affinities with all of the faunal groups identified; however, our material is most comparable with their *Ekwanoscutellum ekwanensis* fauna, from the upper Telychian to ?lowermost Wenlock Chicotte Formation. Their *E. ekwanensis* fauna is predominantly congeneric with that described here, with species of *Ekwanoscutellum, Stenopareia, ?Distyrax, Hyrokybe* and *Scotoharpes* represented.

SYSTEMATIC PALAEONTOLOGY

This published work and the nomenclatural acts it contains, have been registered in Zoobank: http://zoobank.org/References/AA02FA19-080B-413E-936E-5145285D8432

The object of this study was to document the trilobite faunas that occur in the Silurian reefs of North Greenland: 65 different taxa are represented, belonging to 13 families. Given this diversity, and the large number of related species, no phylogenetic analysis of the taxa described has been attempted.

Terminology

Terminology follows Whittington and Kelly (1997) except as outlined below. Holloway and Lane (2012, pp. 417–419, fig. 3) provided a comprehensive review of the homology of lateral glabellar furrows, and cephalic borders and border furrows in the Scutelluidae. They applied the term *articulating flange furrow* to the furrow positioned posteriorly to the posterior border furrow, which separates the posterior articulating flange from the rest of the fixigena, and *bolus*, to define the central swelling or node contained within the posteriormost lateral glabellar furrow. Other terms applied to the Illaenina are as follows: *cusp* employed by Curtis and Lane (1997, p. 14) to describe single or multiple anteriorly directed projections of the anterior margin of the pygidial doublure; *omphalus* employed by Holloway and Lane (1998, pp. 863–864, text-fig. 6) to describe a raised boss, commonly with a median depression, at which the axial furrow may terminate anteriorly; *anterolateral internal pit* employed by Holloway and Lane (1998, p. 864, text-fig. 6) to describe a small pit on the interior of the cranidium (appearing as a node on internal moulds), situated just in front of the omphalus, and usually slightly adaxial or abaxial to it. Proetid terminology includes *intramarginal zone* employed by Owens (2006, p. 119) to describe a flat band between the border and epiborder furrows, as seen in *Winiskia*. Additionally, the term *preocular fixigenal furrow* is introduced here to describe a transversely trending furrow within the fixigena which is adaxially confluent with the preglabellar furrow, as seen in *Airophryus* gen. nov. Terminology used for encrinurids follows that of Edgecombe and Chatterton (1993, p. 77) where PL = preglabellar lateral lobe, referring to the swelling on the lateral part of the anterior border of the cranidium.

Order CORYNEXOCHIDA Kobayashi, 1935
Suborder ILLAENINA Jaanusson *in* Moore, 1959

Remarks. Lane and Thomas (1983) presented a comprehensive history of the classification of the Illaenina. The suborder comprises the families Illaenidae Hupé, 1953, Phillipsinellidae Whittington, 1950, Scutelluidae Richter and Richter, 1955 and Styginidae Vogdes, 1890.

Family ILLAENIDAE Hawle and Corda, 1847

Remarks. Effacement is a condition found commonly among genera of the Illaenina: characteristically, there is a progressive loss of the dorsal furrows, often accompanied by increased exoskeletal convexity and increase in relative axis width (Lane and Thomas 1983). An important consequence of effacement is that it obscures or removes many characters of taxonomic value, making the inference of relationships particularly difficult: the family-level classification of the genera concerned has proved especially contentious. There is general agreement that some effaced genera (collectively termed illaenimorphs), such as *Stenopareia* Holm, 1886, and *Bumastoides* Whittington, 1954, are closely related to *Illaenus* and unequivocally belong to the Illaenidae (Lane and Thomas 1983; Whittington 1997, 1999, 2000; Edgecombe *et al.* 2006). Such genera generally have small eyes, distinct axial furrows may be retained posteriorly in the cephalon and anteriorly in the pygidium, and the thoracic axis remains relatively narrow.

Other highly effaced illaenimorph genera, particularly those traditionally assigned to the Bumastinae Raymond, 1916 – such as *Bumastus* Murchison, 1839 and *Cybantyx* Lane and Thomas *in* Thomas, 1978 – differ from those included unequivocally in the Illaenidae in having a much wider and weakly defined thoracic axis, and a proportionately larger pygidium which exhibits a similar convexity to the cranidium. Typically, the pygidial axis is effaced, and the eye is significantly larger also. Lane and Thomas (1978) regarded such taxa, which had been previously assigned to the Illaenidae, as highly effaced scutelluids.

Genus STENOPAREIA Holm, 1886

Type species. By original designation; *Illaenus linnarssonii* Holm, 1882, from the Ordovician of Dalarne, Sweden.

Other species. Stenopareia acymata (Howells, 1982); *S. aemula* (Salter, 1867); *S. americana* (Raymond, 1916); *S. angulata* Maksimova, 1962; *S. aplata* (Raymond, 1925); *S. balclatchiensis* (Reed, 1904); *S. borealica* (Balashova, 1959); *S. bowmanni* (Salter *in* Phillips and Salter, 1848); *S. camladica* Whittard, 1961; *S. catathema* (Howells, 1982); *S. craigensis* (Reed, 1935); *S. duffyae* Chatterton and Ludvigsen, 2004; *S. gardenensis* (Shaw, 1968); *S. garsonensis* Westrop and Ludvigsen, 1983; *S. glaber* (Kjerulf, 1865); *S. globosa* (Billings, 1859); *S. glochin* (Howells, 1982); *S. grandis* (Billings, 1859); *S. hospes* (Barrande, 1872); *S. illtyd* Ludvigsen and Tripp, 1990; *S. imperator* (Hall, 1861); *S. johnstoni* (Etheridge, 1896); *S. linnarssonii* (Holm, 1882); *S. lissbergensis* (Warburg, 1925); *S. livonica* (Holm, 1886); *S. longicapitata* (Reed, 1896); *S. longispinosa* (Kiær, 1908); *S. marshalli* (Salter, 1867); *S. megacipitis* Ju *in* Qiu *et al.*, 1983; *S. miaopoensis* Lu, 1975; *S. nexilis* (Salter, 1867); *S. norvegica* (Whittard, 1939); *S. oblita* (Barrande, 1872); *S. ophiocephala* (McCoy, 1846); *S. oviformis* (Warburg, 1925); *S. pamirica* (Balashova, 1966); *S. panderi* (Barrande, 1852); *S. perceensis* (Cooper, 1930); *S. persica* sp. nov.; *S. postrema* (Kiær, 1908); *S. proles* (Holm, 1886); *S. pterocephala* (Whitfield, 1878); *S. pulchella* Šnajdr, 1975; *S. pulchrum* Apollonov, 1974; *S. recta* Ancygin, 1973; *S. rivulus* (Ingham and Tripp, 1991); *S. rotunda* (Kiær, 1908); *S. sculpta* (Kiær, 1908); *S. shallochesis* (Reed, 1904); *S. shelvensis* (Whittard, 1938); *S. somnifer* Lane, 1979; *S. transversa* Ju *in* Qiu *et al.*, 1983; *S. trippi* (Morris, 1988); *S. thomsoni* (Salter, 1867); *S. transmota* (Reed, 1935); *S. tyronensis* (Reed, 1933); *S. xintanensis* Xia, 1978; *S.? julli* Norford, 1981.

Remarks. In their diagnosis of the genus, Curtis and Lane (1997, p. 18) stated that the axial furrow is present only behind the lunette, but it may extend farther forwards for a short distance. As with all illaenimorph taxa, the effaced

nature of *Stenopareia* can significantly hamper description and identification of species, but the number of cusps on the pygidial doublure is an apomorphy, as recognized by Jaanusson (1954, p. 574; 1957, p. 109) and Curtis and Lane (1997, p. 19). Cladistic analysis by Carlucci *et al.* (2012) resulted in the transfer of three species from *Bumastoides* to *Stenopareia* (*S. aplatus* (Raymond, 1925), *S. gardenensis* (Shaw, 1968) and *S. rivulus* (Ingham and Tripp, 1991)).

Stenopareia persica sp. nov.
Figures 2, 3A–F

LSID. urn:lsid:zoobank.org:act:AF3FA125-1E5B-4660-8252-C0C121EDD1B2

Derivation of name. Latin, *persica*, peach. Noun in apposition.

Holotype. MGUH30535 (Fig. 2A–D) cranidium; from GGU 275049 (VGL3).

Figured paratypes. Locality CPL5: MGUH30536–30537 cranidia; MGUH30538–30539 librigenae; MGUH30540 rostral plate; MGUH30543 pygidium. Locality VGL1: MGUH30541 pygidium. Locality CPL2: MGUH30542 pygidium.

Diagnosis. Reaching large size, cephalic length may exceed 60 mm; axial furrows converging and then diverging posteriorly behind palpebral lobe in palpebral view; anterior and lateral cephalic margin rim-like; pygidium with well developed articulating facet and a median cusp on inner edge of doublure.

Description. Cranidia measuring from 7.7 to 68.4 mm sagittal length, 59% as high as long (sag.), with maximum height about halfway between palpebral lobe and anterior margin. Cranidium 91% as long (sag.) as wide (tr.) across palpebral lobes (range = 85–97%; n = 6). From there, furrows are subparallel until anterior of lunettes; latter shallow, exsagittally elongated and placed opposite anterior half of palpebral lobe in palpebral view. Glabellar width at lunettes 92% that at posterior margin (range = 85–99%; n = 8). Anterior to lunettes, axial furrows smoothly diverging forwards at roughly 20 degrees to an exsagittal line, fading roughly halfway between palpebral lobe and anterior margin. Lateral glabellar muscle impressions visible as striated areas on internal moulds; S1 long (exsag.), extending from anterior end of lunettes until point where axial furrows start to fade (Fig. 2E, H). Palpebral lobe comprising on average 16% cranidial sagittal length (range = 16–23%; n = 6), and positioned on average 26% cranidial sagittal length, from posterior border (range = 25–29%; n = 5). Sculpture of fairly regularly spaced and

FIG. 2. *Stenopareia persica* sp. nov. A–D, MGUH30535 holotype cranidium, GGU 275049 (VGL3); A, plan, B, palpebral, C, anterior, and D, lateral views. E, H, MGUH30536 cranidium, GGU 274689 (CPL5); E, plan and H, lateral views. F, MGUH30537 cranidium, GGU 274689 (CPL5) plan view. G, J, MGUH30538 librigena, GGU 274689 (CPL5); G, dorsal and J, ventral views. I, MGUH30539 librigena, GGU 274689 (CPL5) dorsal view. K, MGUH30540 rostral plate, GGU 274689 (CPL5) ventral view. All scale bars represent 10 mm.

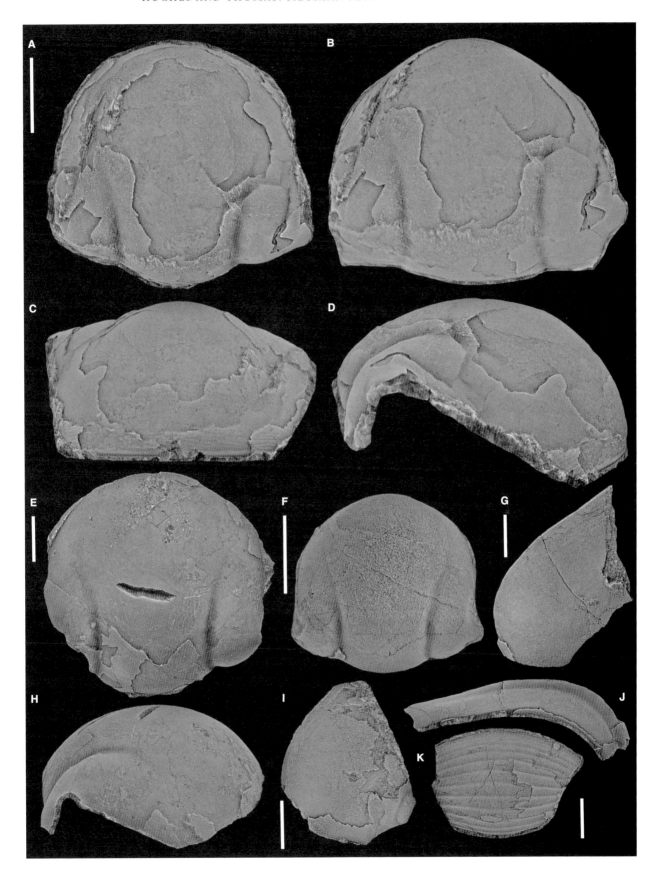

continuous, non-anastomosing and transversely trending terrace ridges approaching anterior margin of cranidium.

Librigena most convex posterolaterally. Lateral margin with rim like that of anterior cranidial border. Genal angle well rounded. Where preservation permits, short, discontinuous terrace ridges seen laterally. Doublure with distinct vincular groove converging with outer margin anteriorly.

Rostral plate most convex (sag., tr.) medially; connective sutures slightly sigmoidal and diverging forwards at about 30 degrees to an exsagittal line. 11 or 12 prominent terrace ridges trending transversely.

Pygidium sagittal lengths measuring from 8.3 to 48.8 mm; about 31% as high as long (sag.) and maximum height about one-third pygidial sagittal length from posterior margin; 68% as long (sag.) as wide (tr.; range = 59–73%; n = 5). Axis comprising 36% (range = 32–39%; n = 5) maximum pygidial width (tr.) its anterior edge evenly convex forwards. Doublure comprising 19% (n = 1) total pygidial sagittal length (excluding cusp); with distinct, continuous terrace ridges.

Ontogeny. The pygidia display a change from a more polygonal outline to a more nearly semicircular one during growth, a pattern described from *S. grandis* (Billings, 1859) by Chatterton and Ludvigsen (2004). The articulating facet in smaller specimens comprises a much greater proportion of pygidial length (Fig. 3E–F), with the abaxial end placed opposite 44% pygidial length from anterior margin medially (sag.), this decreasing to 25% in larger specimens (Fig. 3A–D). This change in proportions is also seen in *S. grandis* (Chatterton and Ludvigsen 2004, p. 17, pl. 1, figs 10–11; pl. 2, figs 4–5, 11–16; pl. 3, figs 5–8) and *S. somnifer* (Lane 1979, p. 14, pl. 1, figs 17, 19–23, 25; pl. 2, figs 10, 13; pl. 3, figs 1–2; pl. 5, fig. 3), making it useful in the identification of juvenile specimens.

Remarks. *Stenopareia persica* is comparable to *S. grandis* in terms of its cranidial and pygidial proportions, and large maximum size; Chatterton and Ludvigsen (2004) noted that *S. grandis* can exceed sagittal lengths of 100 mm. The presence of a single cusp on the inner margin of the pygidial doublure is a synapomorphy. *Stenopareia persica* lacks the socle-like swelling which *S. grandis* exhibits under the middle of its visual surface. Other differences, which additionally distinguish *S. persica* from other well known Silurian species from Laurentia, such as *S. thomsoni* (Salter, 1867) (see Howells 1982, pp. 12–14, pl. 3, figs 1–8), and *S.? julli* Norford, 1981 (pp. 7–8, pl. 6; pl. 10, figs 8–10), include the morphology of the axial furrows which are less effaced anteriorly in *S. persica* and converge anteriorly for a much greater proportion of their course. The lunettes are situated more posteriorly in *S. persica*, and the pygidial axis is proportionally narrower (tr.). The cranidium of *S. persica* is very distinctive for the genus and is readily differentiated from all other species of *Stenopareia* in the possession of a *Cybantyx*-like

upturned rim on the anterior and anterolateral cephalic margin.

Occurrence and distribution. Common in Central Peary Land (CPL2, 4–5), rare in Valdemar Glückstadt Land (VGL1–2).

Stenopareia cf. *grandis* Billings, 1859
Figure 3G–I

Remarks. Cephala, cranidia and pygidia exhibit similarities to *S. grandis* (see Chatterton and Ludvigsen 2004, p. 17, pls 1–2), including their large size. Cephala and cranidia have sagittal lengths ranging from 5.3 to 20 mm. Cranidia are 84% as long (sag.) as wide (tr.) across palpebral lobes (range = 83–85%; n = 3), and a measurable cephalon is 68% as long (sag.) as wide (tr.) at its widest point opposite palpebral lobes. Cephalic similarities with *S. grandis* include the nature of the axial furrows, converging anteriorly from the posterior margin until intersecting with the faint lunettes, anterior of which they are effaced. Additionally, there is a socle-like swelling under the middle of the eye, and the presence of fairly continuous, irregular terrace ridges, trending transversely but bowed medially, and present near the anterior margin and posterior of cephalon. However, in *S.* cf. *grandis* and other Greenland species of *Stenopareia* (*S. persica*; *S. somnifer* Lane, 1979, pp. 14–17, pl. 1, figs 17–25; pl. 2, figs 1–14; pl. 3, figs 1–2; pl. 5, fig. 3), the lunettes are positioned more or less opposite the palpebral lobe, not well anterior of it as in *S. grandis*.

Pygidia have sagittal lengths ranging from 9.3 to 16.3 mm, and are 65% as long (sag.) as wide (tr.) (range = 62–69%; n = 5). One pygidium has a pair of exsagittally directed oval muscle scars situated at the midlength of the articulating facets (Fig. 3I). The pygidial length/width ratios agree with *S. grandis*, as does a median arch with a fairly straight anterior margin. The pygidial axis comprises 38% of the maximum pygidial width (tr.) and is slightly narrower anteriorly than in *S. grandis*. The axial furrows, however, are not as shallow as in *S. grandis*.

The position of the lunettes, and the axial morphology, result in this material being assigned to *S.* cf. *grandis*.

Occurrence and distribution. Most common in Western Peary Land (WPL1–2). Also present in Wulff Land (WL) and South Nares Land (SNL2).

Stenopareia sp.
Figures 3J–K, 4A–C

Remarks. Fragmentary cranidia and pygidia from Valdemar Glückstadt Land are closest to *S.* cf. *grandis* but differ as follows: cranidia are 97% as long (sag.) as wide (tr.) (n = 2) as opposed to 84%; pygidia are 57% as long (sag.) as wide (tr.) (n = 2) as opposed to 65%; pygidial axial furrows are deeper; pygidial terrace ridges are stronger, are strongly bowed forwards medially and become more discontinuous laterally. Given the limited material and lack of other distinguishing characters, the material is kept under open nomenclature.

Occurrence and distribution. Found in Valdemar Glückstadt Land (VGL3).

Family SCUTELLUIDAE Richter and Richter, 1955

Remarks. Lane and Thomas (1983) noted that the phylogenetic relationships between genera assigned to the Styginidae and Scutelluidae were uncertain, but took the general similarities between the taxa concerned to be evidence of close affinities. They therefore regarded the (Ordovician) styginids as plesiomorphic representatives of the Silurian–Devonian Scutelluidae, and assigned all of the genera to an undivided Styginidae. An opposing argument was presented by Ludvigsen and Chatterton (1980), who suggested that the bumastines (their terminology to include such effaced genera as *Bumastus* and *Rhaxeros*) arose from the styginids and that the bumastines later gave rise to the scutelluids. Holloway (2007) also regarded the Styginidae as separate from the Scutelluidae. None of these interpretations was supported by cladistic analysis, however. A potentially important character distinguishing non-effaced styginids and scutelluids is that the furrows separating the radial ribs on the pygidium are pleural in the styginids and interpleural in scutelluids (Holloway 2007). Most recently, Adrain (2013) argued that the separation of styginids, scutelluids and bumastines would result in three groups which are not monophyletic, and followed Lane and Thomas' (1983) recognition of a single family.

An extensive phylogenetic analysis is needed to try to resolve the issues summarized, particularly as it is possible that highly effaced taxa evolved on more than one occasion from different ancestors. Pending such a revision, we provisionally retain the Scutelluidae as a family separate from the Styginidae. What might be termed 'conventional' scutelluids, with impressed dorsal furrows and distinct radial ribs on the pygidium, are assigned to the Subfamily Scutelluinae Richter and Richter, 1955. The more highly effaced taxa are here assigned provisionally to the Subfamily Bumastinae. This segregation into two groups of morphologically similar taxa facilitates description and comparison.

Subfamily SCUTELLUINAE Richter and Richter, 1955

Genus EKWANOSCUTELLUM Pribyl and Vaněk, 1971

Type species. By original designation; *Bronteus ekwanensis* Whiteaves, 1904, from the Telychian – ?lower Wenlock Attawapiskat Formation, Ekwan River, Ontario, Canada.

Other species. Ekwanoscutellum agmen sp. nov.; *E. laphami* (Whitfield, 1882).

Diagnosis. Anterior border and preglabellar furrow only present laterally. Glabella narrowest at cranidial mid-length at which it displays a strong angularity and expands rapidly forwards to twice that width. Occipital impression and S1 large and well impressed, S2 and S3 much weaker and smaller; may not be visible. Librigena terminates in a genal spine comprising over one-fifth total cephalic sagittal length. Pygidium with length subequal to width, and with maximum width in anterior quarter of pygidial sagittal length; vaulted, with axis roughly one-quarter sagittal length of pygidium. Median rib bifid for roughly its posterior half.

Remarks. Whiteaves' original material of *E. ekwanensis* comprised solely pygidia. Norford (1981) refigured these and described much additional material. He noted that the lack of sculpture on much of the dorsal surface of *E. ekwanensis* could be characteristic of the genus. However, our new species *E. agmen* has more widespread terrace ridges dorsally, demonstrating that this is not the case. Sculpture is also diagnostic at species level in *Opoa*; *Opoa limatula* sp. nov. has a very different sculpture from the type species *O. adamsi*.

Ekwanoscutellum shares many similarities with *Meroperix* particularly the overall cranidial shape, the positions of the lateral glabellar muscle impressions, and the lack of a preglabellar furrow and anterior border medially. The most important differences are as follows: S3 is larger than S2 in *Ekwanoscutellum*, whereas the reverse is true for *Meroperix*; the *Meroperix* pygidium is wider as long and that of *Ekwanoscutellum* is about as wide as long; the pygidium of *Ekwanoscutellum* is more convex than that of *Meroperix*, which is virtually flat; the pygidial axis of *Ekwanoscutellum* is as long as or longer than wide and parabolic in outline; that of *Meroperix* is wider than long and subtriangular to bell-shaped in outline; the pygidial axis of *Ekwanoscutellum* comprises one-quarter of the entire pygidial length whereas the pygidial axis of *Meroperix* is proportionally shorter, comprising one-fifth of the entire pygidial length; *Ekwanoscutellum* reaches a larger maximum size than *Meroperix*.

Norford (1981) was not confident in assigning *Bronteus laphami* Whitfield, 1882 (p. 310, pl. 22, figs 1–4), from the Wenlock of Wisconsin, to *Ekwanoscutellum*. He noted

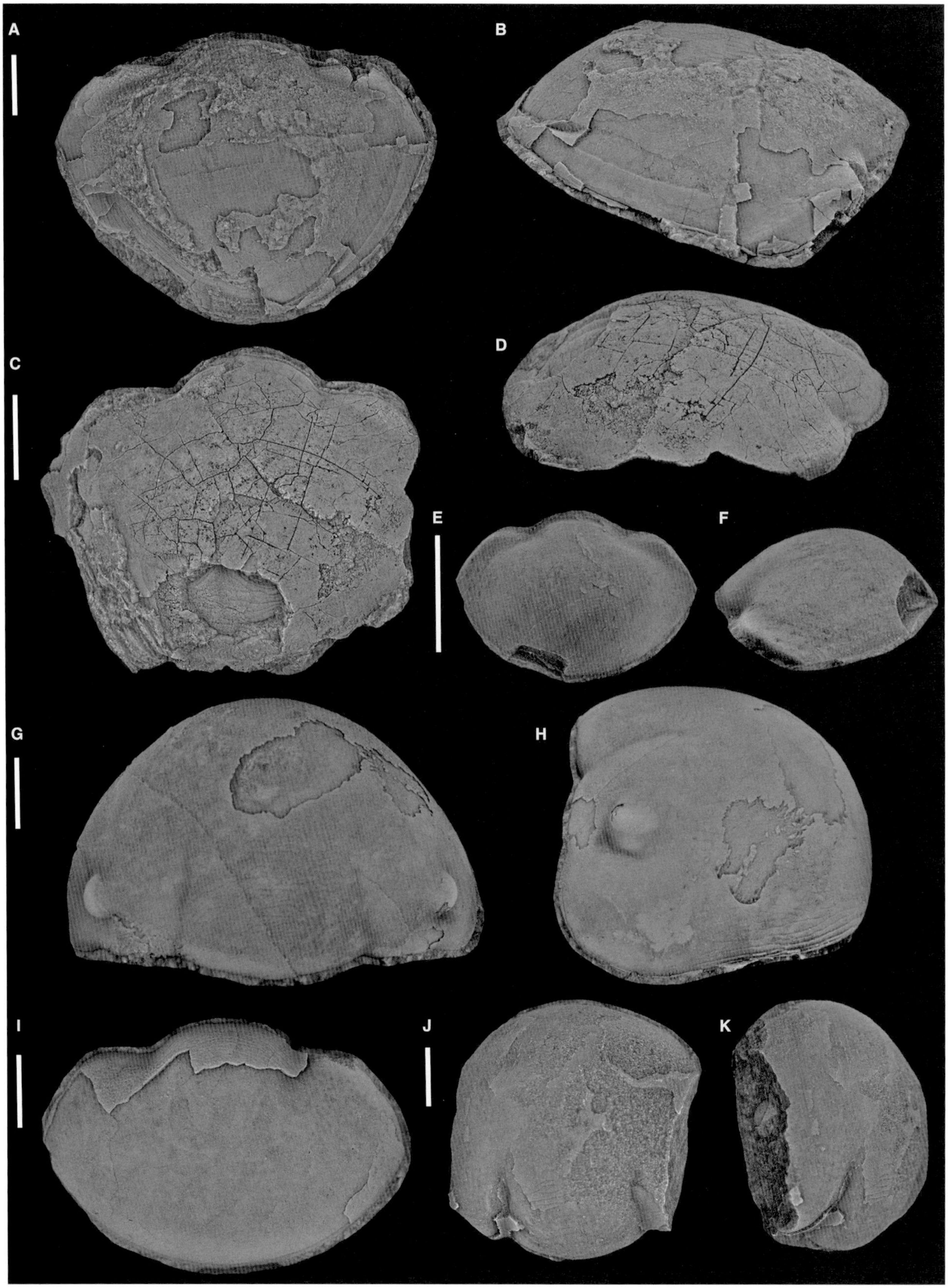
A
B
C
D
E
F
G
H
I
J
K

that the species was essentially described from its pygidium and recognized the importance of cranidial morphology to generic assignment. Whitfield's (1882) illustrations of the pygidia of *B. laphami* do exhibit the characteristic proportions of *Ekwanoscutellum*, and in addition, we have studied the syntypes of *laphami* in the US National Museum of Natural History, Washington, DC: these include a fragmentary cranidium (Fig. 5A) of *B. laphami*, which possesses all the distinguishing characters of *Ekwanoscutellum* – the medial lack of an anterior border and preglabellar furrow, combined with the course of the axial furrows and the effacement of S2 and S3 are all characteristic. A fragmentary pygidium (Fig. 5B) and an external mould of a pygidium (Fig. 5C) confirm proportions which are characteristic of *Ekwanoscutellum*. We can therefore confidently assign *B. laphami* to *Ekwanoscutellum*.

Ekwanoscutellum agmen sp. nov.
Figures 4D–M, 6

LSID. urn:lsid:zoobank.org:act:4E846173-04E9-4560-81C1-80EA28BABA73

Derivation of name. Latin, *agmen*, army on the march, multitude, crowd, train. This is the most abundant species described here. Noun in apposition.

Holotype. MGUH30549 (Fig. 4D–F) cranidium; from GGU 274689 (CPL5).

Figured paratypes. Locality CPL2: MGUH30551, MGUH30555 cranidia; MGUH30559 librigena; MGUH30561 rostral plate; MGUH30565 thoracic segment; MGUH30568–30569, MGUH30571–30572 pygidia. Locality CPL4: MGUH30562 hypostome; MGUH30566 pygidium. Locality CPL5: MGUH30553 cranidium; MGUH30557–30558 librigenae; MGUH30560 rostral plate; MGUH30564 thoracic segment; MGUH30570 pygidium. Locality VGL1: MGUH30552, MGUH30554 cranidia. Locality KCL1: MGUH30550, MGUH30556 cranidia; MGUH30563 hypostome; MGUH30573 pygidium. Locality KCL3: MGUH30567 pygidium.

Diagnosis. Fixigena relatively narrow (tr.) anteriorly; both pygidium and pygidial axis as wide as or a little wider (tr.) than long (sag.); pygidial maximum width (tr.) located opposite posteriormost half of axis; pleural region slightly concave towards margin.

Description. Cranidium approximately equal in width at β–β and ω–ω, a little narrower at δ–δ; width at γ–γ roughly equal to

length (sag.). Maximum height at L1 in lateral view. Axial furrows well impressed, parallel in their posterior halves except for a slight inflexion opposite lunettes. Axial furrows abruptly change course at a point 50% total cranidial sagittal length from posterior (range = 45–53%; n = 16), coincident with the anterior end of S1, where they diverge strongly at 35–50 degrees (n = 8) continuing in a straight line to intersect with the anterolateral border furrow. There, glabella about twice as wide as at narrowest part (range = 2.1–2.5 times; n = 11). Occipital ring approximately twice as long sagittally as exsagittally, sagittal length about one-third width (range = 30–37%; n = 14). Well impressed occipital furrow and oval occipital impressions connected to axial furrow. Narrow anterior border and preglabellar furrow best developed laterally to point nearly in line (exsag.) with posterior portion of axial furrows, fading medially. S1 exsagittally oval with a broad, indistinct swelling, connected to axial furrow. S2 and S3 effaced, rarely visible, situated successively farther from sagittal line and close to but not connected to axial furrows (Fig. 4L); S2 smallest, transversely oval, separated from but close to S1; S3 approaching size of S1 but weaker, the least well impressed of the muscle impressions, transversely oval. Sculpture comprising terrace ridges; covering anterior of glabella and anterior border where prominent, continuous, non-anastomosing, roughly parallel to anterior margin. Fainter, transversely trending terrace ridges on fixigena. Largest cranidium is 30 mm long.

Fixigena gently convex, posterior edge more or less opposite median section of occipital furrow. Palpebral lobe at ε opposite anterior part of occipital impressions, at δ–δ located 30% total cranidial sagittal length from posterior border (range = 28–32%; n = 10), extending forwards to middle (exsag.) of S1. Lunette oval, placed between anterior of occipital muscle impression and posterior of S1. Eye ridge intersecting axial furrow transversely opposite L3. Anterolateral concave zone longer (exsag.) than anterolateral border.

Librigena with large, kidney-shaped and strongly raised visual surface. Epiborder furrow deepest anteriorly, becoming broader and more diffuse posteriorly and extending weakly onto genal spine (Fig. 6A–B). Lateral border furrow well impressed, converging towards epiborder furrow anteriorly. Transversely trending, persistent, regularly spaced terrace ridges which locally anastomose and become increasingly prominent posteriorly. Doublure with terrace ridges mirroring direction of epiborder and lateral border furrows (Fig. 6C).

Rostral plate with connective sutures diverging forwards at 50–60 degrees to sagittal line, slightly curved. Prominent, continuous terrace ridges trend transversely over entire rostral plate (Fig. 4D–E).

Hypostome with poorly developed shoulders. Lenticular in outline, posterior margin terminating in blunted point. Short posterior lobe of middle body, prominent maculae. Total sagittal length 62% of width (tr.) across anterior wings (range = 57–67%;

FIG. 3. A–F, *Stenopareia persica* sp. nov.; A–B, MGUH30541 pygidium, GGU 275038 (VGL1); A, dorsal and B, ventral views; C–D, MGUH30542 pygidium, GGU 198136 (CPL2); C, dorsal and D, oblique lateral views; E–F, MGUH30543 pygidium, GGU 274689 (CPL5); E, dorsal and F, lateral views. G–I, *Stenopareia* cf. *grandis* Billings, 1859; G–H, MGUH30544 cephalon, GGU 298506 (WL); G, dorsal and H, lateral views; I, MGUH30545 pygidium, GGU 301318 (WPL1) dorsal view. J–K, *Stenopareia* sp. MGUH30546 cranidium, GGU 275049 (VGL3); J, dorsal and K, oblique lateral views. Scale bars represent 10 mm (A–D) and 5 mm (E–K).

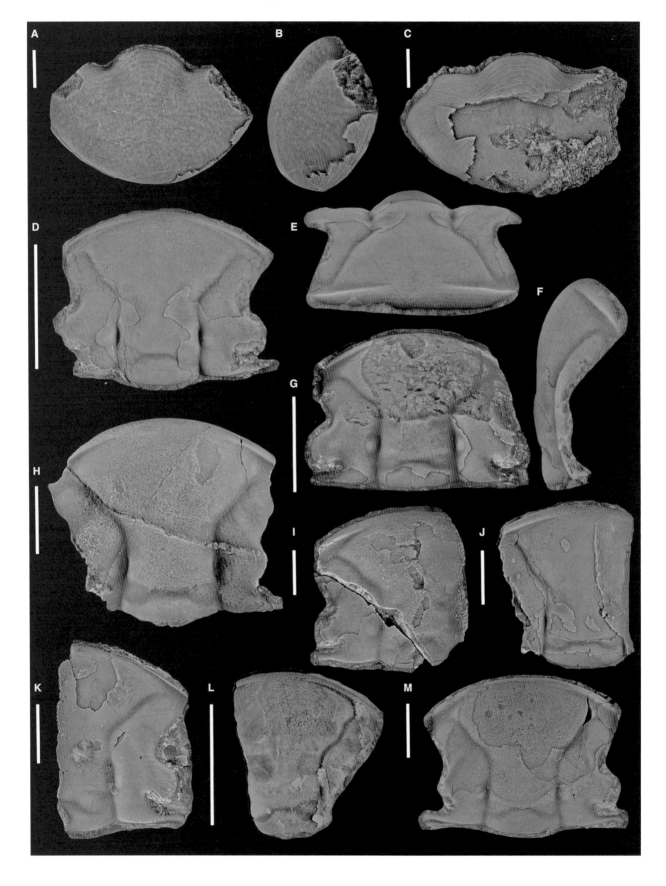

n = 4). Prominent terrace ridges cover entire hypostome, trending in a 'U shape'. Thoracic segments all disarticulated; covered in transversely trending continuous terrace ridges. Articulating half ring roughly half length (sag.) of axial ring; well impressed axial furrow. Both axial ring and articulating half ring convex; pleural region flatter with obliquely trending terrace ridges.

Pygidium an anteriorly truncated ellipse, averaging 86% as long (sag.) as wide (tr.) (range = 77–96%; n = 17); when plotted, the pygidial transverse width = 1.08 pygidial sagittal length +1.38 (see Remarks for discussion, and Fig. 7A). Axis varying from slightly wider than long to slightly longer than wide, averaging 86% as long (sag.) as wide (tr.) (range = 63–103%; n = 38); when plotted, the pygidial axis transverse width = 1.15 pygidial axis sagittal length +0.10 (see Remarks for discussion, and Fig. 7B). Axis comprising about one-quarter maximum pygidial width anteriorly, parabolic to very weakly bell-shaped in outline, with maximum convexity anteriorly. Very well impressed axial furrows. Axis constitutes 24% total pygidial sagittal length (range = 21–30%; n = 16); when plotted, the pygidial axis sagittal length = 0.22 pygidial sagittal length +0.38 (see Remarks for discussion, and Fig. 7C). Adaxial half of pleural region gently convex. Pleural ribs slightly sigmoidal. Median rib always bifid for at least one-third of its length (sag.); twice as wide distally as adjacent paired ribs. All pleural ribs transversely convex. Interpleural furrows well impressed proximally, becoming fainter distally; both furrows and ribs gradually widen marginally. Sculpture of faint to prominent non-anastomosing terrace ridges, trending transversely across ribs; abaxially on axis they are deflected anteriorly, trending transversely. Doublure extending just over half pygidial sagittal length, with pronounced ribs and furrows mirroring the dorsal surface. Well developed terrace ridges become increasingly scalloped towards margins (Fig. 4S). Largest pygidium is 68 mm long.

Remarks. Specimens of *E. agmen* display the following intraspecific variation: the medial extent of the anterior border and preglabellar furrow (one large cranidium from Kronprins Christian Land (MGUH30556; Fig. 4M) is identical to other specimens in all respects other than possessing a complete anterior border); the anterior course of the axial furrows ranges from straight (Fig. 4H) to sigmoidal (Fig. 4L); the pygidial axis may be straight-sided (Fig. 6Q) or weakly convex outward (Fig. 6K) or 'pinched' inwards (Fig. 6R); and the degree of marginal concavity to the pygidia varies (compare Fig. 6L with 6N).

The course of the cranidial axial furrows, proportion of the anterior margin with an anterior border and preglabellar furrow, the anterior position of the maximum width of the pygidia, the shape and relative length of the pygidial axis, and the morphology of the pygidial ribs and furrows make this species most similar to *E. ekwanensis*. The main differences from that species are as follows: the anterior portion of the fixigena is relatively narrower (tr.); the median part of the occipital ring is a little longer (sag.); the pygidium is a little shorter relative to its width; the lateral pygidial margins straighter, maintaining maximum width for longer; the pygidial axial furrows not as straight as those of *E. ekwanensis*.

Reduced major axis regression has been used previously to estimate the sagittal length of *Ekwanoscutellum* pygidia when, due to fragmentation, only the sagittal length of the axis is measurable (Westrop and Rudkin 1999). Plotting pygidial axial length against pygidial sagittal length, which have a linear relationship, gives a reasonable estimation of total pygidial length. Bivariate plots showing reduced major axis regression have additionally been shown to enable quantitative morphological comparisons between trilobite species and to observe size-related variation. This methodology has been successfully applied to agnostid cephala and pygidia (Westrop and Eoff 2012) and the menomoniid genus *Hysteropleura* (Westrop and Ludvigsen 2000), for example. We have applied reduced major axis regression to the three commonest scutelluid species described here: *E. agmen*; *Jungosulcus emilyae* and *O. magnifica* (Fig. 7). Pygidial length was plotted against pygidial width, pygidial axial length against pygidial axial width and pygidial axial length against total pygidial length. The pygidia of *E. agmen* and *O. magnifica* are superficially similar and may be confused. The application of reduced major axis regression aids comparison between the pygidia of these species: although the straight-line equations for axial width against axial length are extremely similar, those for pygidial width against pygidial length clearly distinguish *E. agmen* from *O. magnifica* (Fig. 7A). The plots of pygidial axial length against pygidial length are extremely similar for *E. agmen*, *O. magnifica* and *J. emilyae* despite the obvious differences in overall pygidial morphology between the latter two of these species, demonstrating that this proportion is not a useful character state in distinguishing these taxa.

Occurrence and distribution. Abundant in Central Peary Land (CPL1–5), also present in Valdemar Glückstadt Land (VGL1) and Kronprins Christian Land (KCL1, 3, 5).

FIG. 4. A–C, *Stenopareia* sp.; A–B, MGUH30547 pygidium, GGU 275049 (VGL3); A, dorsal and B, oblique lateral views; C, MGUH30548, pygidium, GGU 275049 (VGL3) dorsal view. D–M, *Ekwanoscutellum agmen* sp. nov.; D–F, MGUH30549 holotype cranidium, GGU 274689 (CPL5); D, dorsal, E, dorsoanterior and F, lateral views; G, MGUH30550 cranidium, GGU 275015 (KCL1) dorsal view; H, MGUH30551 cranidium, GGU 198186 (CPL2) dorsal view; I, MGUH30552 cranidium, GGU 275038 (VGL1) dorsal view; J, MGUH30553 cranidium, GGU 274689 (CPL5) dorsal view; K, MGUH30554 cranidium, GGU 275038 (VGL1) dorsal view; L, MGUH30555 cranidium (not blackened), GGU 198216 (CPL2) dorsal view; M, MGUH30556 cranidium, GGU 275015 (KCL1) dorsal view. Scale bars represent 5 mm (A–C) and 10 mm (D–M).

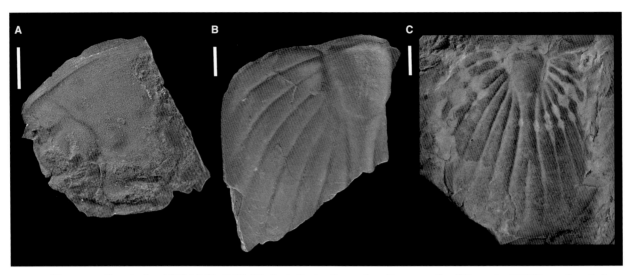

FIG. 5. *Ekwanoscutellum laphami* (Whitfield, 1882). Wenlock, Racine Formation, Kewaunee Co., Wisconsin. USNM numbers indicate specimens belong to the type and figured collection, Smithsonian Institute, Washington. A, fragmentary cranidium, USNM 135947. B, fragmentary pygidium, USNM 135953. C, pygidial negative, USNM 137042. All are syntypes. All scale bars represent 10 mm.

Genus JUNGOSULCUS gen. nov.

LSID. urn:lsid:zoobank.org:act:F9F1F748-75D0-4168-8800-ADB7CE41BA71

Derivation of name. Combination of Latin, *jungo*, unite or connect, and Latin, *sulcus*, furrow or groove. Reference to the proximal merging of pygidial ribs five, six and seven, and their associated interpleural furrows. Gender masculine.

Type species. Jungosulcus emilyae sp. nov. from the Telychian Samuelsen Høj Formation, Central Peary Land, North Greenland.

Other species. Jungosulcus anaphalantos sp. nov.; *J. ventricosus* sp. nov.

Diagnosis. Glabella expanding anteriorly from front of L1; preglabellar furrow not defined, lateral extremity of frontal lobe merging with lateral border. S1 semi-elliptical in outline and only distinctly impressed in posterior half, thus appearing arcuate in form. S1 encloses a large, weak central bolus. S2 situated on the anteromedial edge of S1. Pygidium about two-thirds as long as wide; axis almost twice as wide anteriorly as long (sag.), with clear longitu-

dinal trilobation in larger specimens; median rib bifid or with median carina, or both; lateral pygidial ribs five, six and seven merge prior to intersection with axial furrow.

Remarks. The morphology and position of the lateral glabellar muscle impressions, and the nature of the lateral extremity of the frontal lobe – merging with the lateral border – distinguish *Jungosulcus* from all other scutelluids. With respect to cranidial morphology, *Jungosulcus* shares most similarities with *Kosovopeltis* and *Decoroscutellum*. The cranidium of the type species of *Kosovopeltis*, *K. svobodai* Šnajdr, 1958, figured by Šnajdr (1960, pl. 2, figs 3–6; pl. 3, figs 1–16, 18–26; pl. 4, figs 1–2), and of the type species of *Decoroscutellum*, *D. haidingeri* Šnajdr, 1958, figured by Šnajdr 1960 (pl. 5, figs 10–12, 14; pl. 6, figs 5–12), both from the Ludlow Kopanina Formation, Czech Republic, both exhibit a similar anterior marginal shape, and course of the axial furrows to *Jungosulcus*. The most obvious difference between the genera is the shape of the lateral glabellar muscle impressions, with S2 and S3 being more transversely elongate in *Kosovopeltis* and in *Decoroscutellum* than in *Jungosulcus*. The pygidia of *J. emilyae* and *K. svobodai* are similar also, especially in

FIG. 6. *Ekwanoscutellum agmen* sp. nov.; A, MGUH30557 librigena, GGU 274689 (CPL5) dorsal view. B, MGUH30558 librigena, GGU 274689 (CPL5) dorsal view. C, MGUH30559 librigena, GGU 198186 (CPL2) dorsal view. D, MGUH30560 rostral plate, GGU 274689 (CPL5) ventral view. E, MGUH30561 rostral plate, GGU 198136 (CPL2) ventral view. F, MGUH30562 hypostome, GGU 274654 (CPL4) ventral view. G–H, MGUH30563 hypostome, GGU 275015 (KCL1); G, ventral and H, oblique lateral views. I, MGUH30564 thoracic segment, GGU 274689 (CPL5) dorsal view. J, MGUH30565 thoracic segment, GGU 198194 (CPL2) dorsal view. K–L, MGUH30566 pygidium, GGU 274654 (CPL4); K, dorsal and L, lateral views. M–N, MGUH30567 pygidium, GGU 275021 (KCL3); M, dorsal and N, lateral views. O, MGUH30568 pygidium, GGU 198216 (CPL2) dorsal view. P, MGUH30569 pygidium with abnormal fifth pleural rib, GGU 198204 (CPL2) dorsal view. Q, MGUH30570 pygidium, GGU 274689 (CPL5) dorsal view. R, MGUH30571 latex cast of pygidium, GGU 198216 (CPL2) dorsal view. S, MGUH30572 pygidium, GGU 198216 (CPL2) dorsal view. T, MGUH30573 pygidium, GGU 275015 (KCL1) dorsal view. All scale bars represent 10 mm.

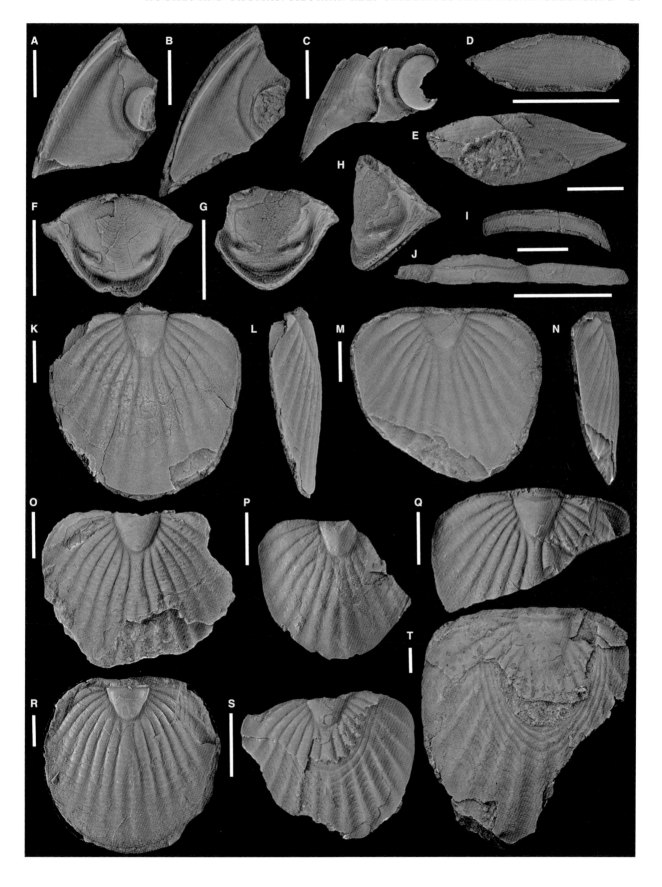

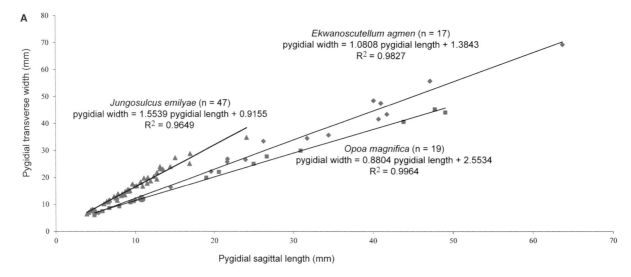

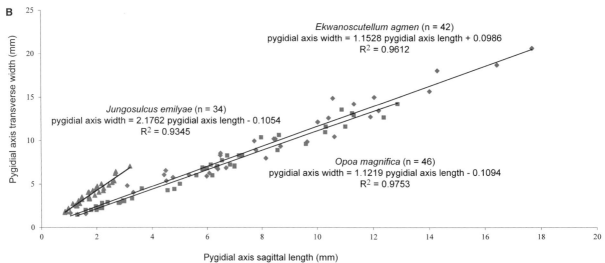

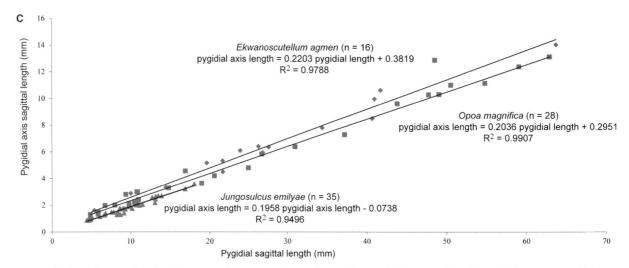

FIG. 7. Major axis regression for *Ekwanoscutellum agmen*, *Jungosulcus emilyae* and *Opoa magnifica*. A, pygidial transverse widths against pygidial sagittal lengths. B, pygidial axial transverse widths against pygidial axial sagittal lengths. C, pygidial axial sagittal lengths against pygidial sagittal lengths.

their outline and form of the pygidial axis. In *K. svobodai*, the pygidial axis is proportionately longer, and the median rib lacks a clear bifurcation. The shallowing of the interpleural furrows resulting in the adaxial merging of pygidial ribs five, six and seven before intersecting the pygidial axis, can be seen too in: *Kosovopeltis insularis* Billings, 1866, as illustrated by Chatterton and Ludvigsen (2004, pl. 5, fig. 8); *Planiscutellum* and some *Kosovopeltis* species illustrated in Šnajdr (1960); *Kosovopeltis allaarti* of Lane (1984, pl. 1, fig. 16) – which was transferred to *Japonoscutellum* by Holloway and Lane (2012, table 2, p. 421) – where it is generally associated with a slight doming of the area and slight forward deflection of the interpleural furrows.

Jungosulcus emilyae sp. nov.
Figures 8, 9A

LSID. urn:lsid:zoobank.org:act:BD39280F-525C-4F2A-8C81-ED3BB2AFC438

Derivation of name. For Emily Thomas, who died suddenly while this work was in progress.

Holotype. MGUH30574 (Fig. 8A–C) cranidium; from GGU locality 274689 (CPL5).

Figured paratypes. Locality CPL5: MGUH30575–30578 cranidia; MGUH30579 eye; MGUH30580 librigena; MGUH30581–30584 pygidia.

Diagnosis. S1–S3 distinctly impressed; pronounced anterolateral border; cranidial axial furrows shallowing anteriorly, terminating as shallow impressions at fixigenal equivalent of lateral border furrow; occipital impressions not connected to axial furrow; median pygidial rib bifid for at least the posterior quarter of its length (sag.).

Description. Cranidium of approximately equal width posteriorly between fulcral sockets as anteriorly between anterolateral margins; width across β–β and δ–δ roughly equal. Width (tr.) across γ–γ subequal to cranidial sagittal length, 75% as long (sag.) as wide (tr.) across δ–δ (range = 67–85%; n = 20). Maximum height in glabella at level of S3. From posterior margin, axial furrows converge anteriorly for a very short distance (until a point just posterior of S1, at which the glabella is at its narrowest); lunettes positioned there. Axial furrows diverge smoothly forwards; remaining well impressed until they become shallower between S2 and S3. Axial furrows terminating at fixigenal extension of the lateral border furrow; glabella widest there; 2.4 times wider (tr.) than at its narrowest (range = 2.1–2.8 times; n = 6). S1 adjacent to axial furrow; shallowest and largest of lateral glabellar muscle impressions. S2 less than half the size of S1, subcircular and very deeply impressed; located

further from axial furrow than S1. S3 similar in size, shape and depth of impression to S2; located at same distance from S2 as S2 is from S1 and a similar distance from axial furrow as S2. Anterior cephalic margin raised laterally forming a prominent anterolateral border (Fig. 8A–C); medially, this is replaced by transversely trending, continuous terrace ridges. Posterior to shallow anterior border furrow is a concave zone longer (exsag.) than anterolateral border and bounded posteriorly by a deeper furrow which corresponds to the lateral border furrow. Occipital ring longest medially where almost twice as long (sag.) as abaxially; comprising 20% total cranidial sagittal length (range = 14–24%; n = 21). Occipital furrow broad and deeply impressed, abaxially terminating as subcircular occipital impressions.

Fixigena with raised palpebral lobe forming highest point. Lunettes oval, located in posterior half of palpebral lobe. Eye ridge meeting axial furrow between S2 and S3. Long anterolateral border furrow.

Librigena subtriangular and largely flat. Epiborder furrow becoming deeper and wider towards anterior of librigena. Lateral border furrow shallower and less distinct than epiborder furrow, sweeping around visual surface.

Pygidium an anteriorly truncated ellipse; 61% as long (sag.) as wide (tr.) (range = 55–69%; n = 47); when plotted, the pygidial transverse width = 1.55 pygidial sagittal length +0.92 (see Remarks for discussion, and Fig. 7A) with maximum width very close to anterior margin, about mid-length (exsag.) of first pleural rib. Axis triangular, terminating posteriorly in a sharply pointed tip; 48% as long (sag.) as wide (tr.) (range = 42–54%; n = 34); when plotted, the pygidial axis transverse width = 2.18 pygidial axis sagittal length −0.11 (see Remarks for discussion, and Fig. 7B) comprising 19% total sagittal length of pygidium (range = 15–21%; n = 35); when plotted, the pygidial axis sagittal length = 0.20 pygidial sagittal length −0.07 (see Remarks for *E. agmen* for discussion, and Fig. 7C); well impressed axial furrows. Pleural region gently vaulted in adaxial half, concave marginally. Median rib always bifid for at least one-quarter of its length (sag.), but amount of bifurcation is variable: where bifurcation is longer, it is less distinct anteriorly (Fig. 8K). Faint median carina may be present anterior of bifurcation (Fig. 8M). Median rib twice width of pleural ribs where it reaches posterior margin. Clearly impressed and almost straight interpleural furrows; both furrows and ribs widen towards margins, furrows deeper anteriorly. Ribs with low transverse convexity and furrows terminating very close to margin or reaching it. Pleural region with closely spaced terrace ridges trending transversely across each rib. Doublure very broad, almost three-quarters total pygidial sagittal length; ribs and furrows mirror those on ventral surface and bear persistent terrace ridges which become increasingly scalloped marginally. Maximum pygidial sagittal length measured 34.5 mm.

Remarks. See under remarks for *J. anaphalantos* and *J. ventricosus*.

Occurrence and distribution. Confined to Central Peary Land (CPL1, CPL3).

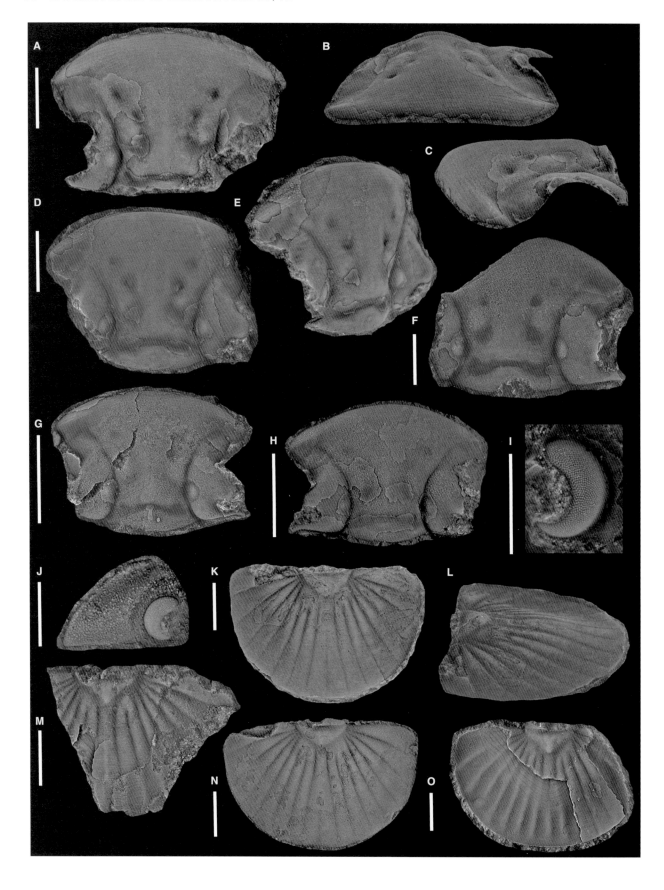

Jungosulcus anaphalantos sp. nov.
Figures 9B, 10A–L, Q

LSID. urn:lsid:zoobank.org:act:49F3AB99-E1E9-4423-B298-8C1DA2A07749

Derivation of name. Greek, *anaphalantos*, with bald forehead, alluding to the gradual effacement of the anteriormost lateral glabellar muscle impressions with growth.

Holotype. MGUH30585 (Fig. 10A–C) cranidium; from GGU locality 298930 (SNL1).

Figured paratypes. Locality SNL1: MGUH30586–30588 cranidia; MGUH30589 librigena; MGUH30590 hypostome; MGUH30591–30594 pygidia.

Diagnosis. Anterior lateral glabellar muscle impressions progressively effaced with growth; cranidial axial furrows terminating at anterolateral border furrow; occipital impressions not connected to axial furrow; median pygidial rib only bifid on smaller specimens, replaced by carina in larger ones.

Description. Cranidium subelliptical, a little wider (tr.) posteriorly between fulcral sockets as anteriorly between anterolateral margins. Width across β–β and δ–δ roughly equal; width (tr.) across γ–γ subequal to cranidial sagittal length. Cranidium 82% as long (sag.) as wide (tr.) across palpebral lobes (δ–δ), (range = 71–92%; n = 11); maximum height of glabella at level of S3. From posterior margin, axial furrows converge anteriorly until just posterior of S1 where lunettes positioned. Axial furrows continue anteriorly with a smooth trajectory; diverging forwards remaining well impressed until they become shallower between S2 and S3; intersecting anterolateral border, where they have a similar depth of impression to anterolateral border. Glabella almost twice as wide (tr.) there as at its narrowest point (average = 1.9 times; range = 1.7–2.0; n = 12). S1 adjacent to axial furrow; most deeply impressed and largest of lateral glabellar muscle impressions. S2 an obliquely directed ellipse; S3 the smallest impression, a transversely directed ellipse. S2 located roughly equidistant between S1 and S3, and is furthest from the axial furrow. Anterior cephalic margin not significantly raised, bounded by moderately impressed anterolateral border furrow. Shallow furrow present just anterior of eye ridge, corresponding to lateral border furrow. Occipital ring with lower convexity than glabella and longest medially where almost twice as long (sag.) as abaxially; comprising 21% total cranidial sagittal length (range = 18–24%; n = 14). Occipital furrow broad, shallower than axial furrow; abaxially terminating at subcircular occipital impressions.

Raised palpebral lobe forming highest point of fixigena; lunettes oval, weakly impressed, located opposite posterior half of palpebral lobe. Eye ridge meeting axial furrow between S2 and S3.

Cranidial sculpture strongest anterior of anterolateral border furrows, there comprising transversely trending, fairly continuous terrace ridges; weaker, less continuous, and more closely spaced transversely trending terrace ridges present elsewhere on cranidia; these are strongest on occipital ring where they are medially bowed anteriorly. Some pitting evident on fixigena.

Librigena subtriangular and of low convexity. Deepest furrow bounding eye; epiborder and lateral border furrows very shallow. Lateral margins slightly raised, most so anteriorly, forming lateral border. Sculpture of transversely trending, fairly continuous terrace ridges.

Hypostome of typical shield-shaped scutelluid type, 78% as long (sag.) as wide (tr.) across anterior wings. Anterior lobe of middle body subcircular; posterior lobe short (sag., exsag.), not inflated. Prominent maculae bounded by deep middle furrow; better impressed than border furrows. Posterior border longest (sag.) adaxially; lateral borders narrower (tr.) than this behind shoulders. Hypostome except for maculae covered in pronounced, continuous terrace ridges which trend roughly transversely across middle body, and run parallel with border furrows on lateral and posterior borders.

Pygidium very similar to that of *J. emilyae* so that description is best achieved by supplying measurements to compare proportions and noting points of difference. Pygidium 64% as long (sag.) as wide (tr.) (range = 60–69%; n = 12); axis 44% as long (sag.) as wide (tr.) (range = 38–50%; n = 24); axis comprising 18% total sagittal length of pygidia (range = 16–21%; n = 12); only smaller specimens have a median rib which is bifid, and this is just posteriorly and for a small proportion of its length (Fig. 10J). Median carina is always present. Maximum pygidial sagittal length measured 17.5 mm.

Ontogeny. *Jungosulcus anaphalantos* exhibits gradual effacement of S3 with increasing size (compare Figs 10D and 10E), accompanied by loss of bifurcation of the median pygidial rib (compare Figs 10J and 10L).

Remarks. *Jungosulcus anaphalantos* is close to *J. emilyae*, especially in pygidial morphology, the pygidium differing mainly in having a less strongly expanding median rib and in losing bifurcation of the median rib with growth. The principal distinguishing characters all relate to the cephalon (Fig. 9 contains line drawings of the cranidia of the two species for comparison) with *J. anaphalantos* differing from *J. emilyae* as follows: the cranidium is longer (tr.) relative to its width (tr.) and less expanded

FIG. 8. *Jungosulcus emilyae* gen. et sp. nov. A–C, MGUH30574 holotype cranidium; A, dorsal, B, dorsoanterior and C, lateral views. D–E, MGUH30575 cranidium; D, dorsal and E, oblique lateral views. F, MGUH30576 cranidium, dorsal view. G, MGUH30577 cranidium, dorsal view. H, MGUH30578 cranidium, dorsal view. I, MGUH30579 close up of eye, dorsal view. J, MGUH30580 librigena, dorsal view. K–L, MGUH30581 pygidium; K, dorsal and L, oblique dorsolateral views. M, MGUH30582 pygidium, dorsal view. N, MGUH30583 pygidium, dorsal view. O, MGUH30584 pygidium, dorsal view. All material from GGU 274689 (CPL5). Scale bars represent 5 mm (A–H, K–O), 0.5 mm (I) and 1 mm (J).

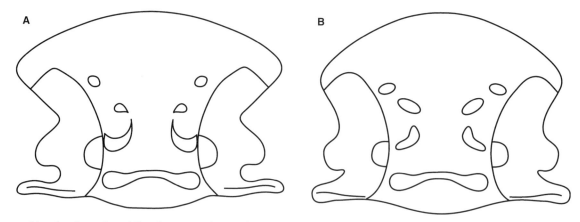

FIG. 9. Line drawings of cranidia of A, *Jungosulcus emilyae* and B, *Jungosulcus anaphalantos*, highlighting morphological differences between the two species.

(tr.) anteriorly; the axial furrows clearly intersect the anterolateral border furrow and do not fade before meeting it as in *J. emilyae*; the glabella is less expanded anteriorly; the lateral glabellar muscle impressions shallow anteriorly, whereas the reverse is true for *J. emilyae*; and S1 and S2 are shallower. In these respects, *J. emilyae* seems more derived than *J. anaphalantos*.

Occurrence and distribution. Confined to South Nares Land (SNL1).

Jungosulcus ventricosus sp. nov.
Figure 10M–U

LSID. urn:lsid:zoobank.org:act:AFCB0855-9535-4D7A-8080-2AC7AE121FEC

Derivation of name. Latin, *ventricosus,* potbellied, bulging. This is the most convex species of *Jungosulcus*.

Holotype. MGUH30595 (Fig. 10M) cranidium; from GGU locality 275038 (VGL1).

Figured paratypes. Locality VGL1: MGUH30596 cranidium; MGUH30597 librigena; MGUH30598–30599 pygidia.

Diagnosis. Convex cranidium and pygidium; S2 and S3 effaced; axial furrows maintaining depth of impression throughout most of course; occipital furrows well impressed, connected to axial furrows; pygidial median rib proximally merged with lateral pleural ribs five, six and seven.

Description. Cranidium subelliptical; slightly wider (tr.) posteriorly between fulcral sockets than anteriorly between anterolateral margins, width across β–β and δ–δ roughly equal. Width (tr.) across γ–γ about equal to cranidial sagittal length. Cranidium just over four-fifths as long (sag.) as wide (tr.) across palpebral lobes at δ–δ. Maximum height in anterior portion of glabella. Axial furrows well impressed, terminating at short (exsag.) anterolateral concave zone; anterolateral border furrow not clearly demarcated. From posterior margin, axial furrows converge anteriorly until intersection with lunettes (roughly two-thirds total cranidial sagittal length from anterior margin) where they are deflected adaxially; anterior of lunettes, diverging anteriorly with a smooth trajectory and maintaining depth of impression. Glabella roughly twice as wide (tr.) anteriorly as at narrowest point. S1 close to but not connected to axial furrow; moderately impressed, bowl-shaped under median node (Fig. 10M–N). Anterolateral border shorter (exsag.) than anterolateral concave zone; anterior cranidial margins not raised medially. Terrace ridges present for roughly most anterior 20% of total cranidial sagittal length. Terrace ridges strongest and continuous anteriorly, becoming more broken and then fading posteriorly. Occipital ring convex, longest sagittally where comprises 18% (n = 2) total cranidial sagittal length. Occipital node at midpoint of occipital ring length (sag.). Occipital furrows abaxially widening at occipital impressions. Occipital ring bears transversely trending, irregular and discontinuous terrace ridges.

Lunettes subcircular, extending from posterior of palpebral lobe, shorter (exsag.) than palpebral lobe. Eye ridge strong. Reg-

FIG. 10. A–L, Q, *Jungosulcus anaphalantos* gen. et sp. nov.; A–C, MGUH30585 holotype cranidium; A, dorsal, B, lateral and C, dorsoanterior views; D, MGUH30586 cranidium, dorsal view; E, MGUH30587 cranidium, dorsal view; F, MGUH30588 cranidium, dorsal view; G, MGUH30589 librigena, dorsal view; H, MGUH30590 hypostome, ventral view; I, MGUH30591 pygidium, dorsal view; J, MGUH30592 pygidium, dorsal view; K, MGUH30593 pygidium, dorsal view; L, Q, MGUH30594 pygidium; L, dorsal and Q, oblique lateral views. All material from GGU 298930 (SNL1). M–U, *Jungosulcus ventricosus* gen. et sp. nov.; M, MGUH30595 holotype cranidium, dorsal view; N–P, MGUH30596 cranidium; N, dorsal, O, lateral and P, dorsoanterior views; R, MGUH30597 librigena, dorsal view; S–T, MGUH30598 pygidium; S, dorsal and T, lateral views; U, MGUH30599 pygidium, dorsal view. All material from GGU 275038 (VGL1). All scale bars represent 5 mm.

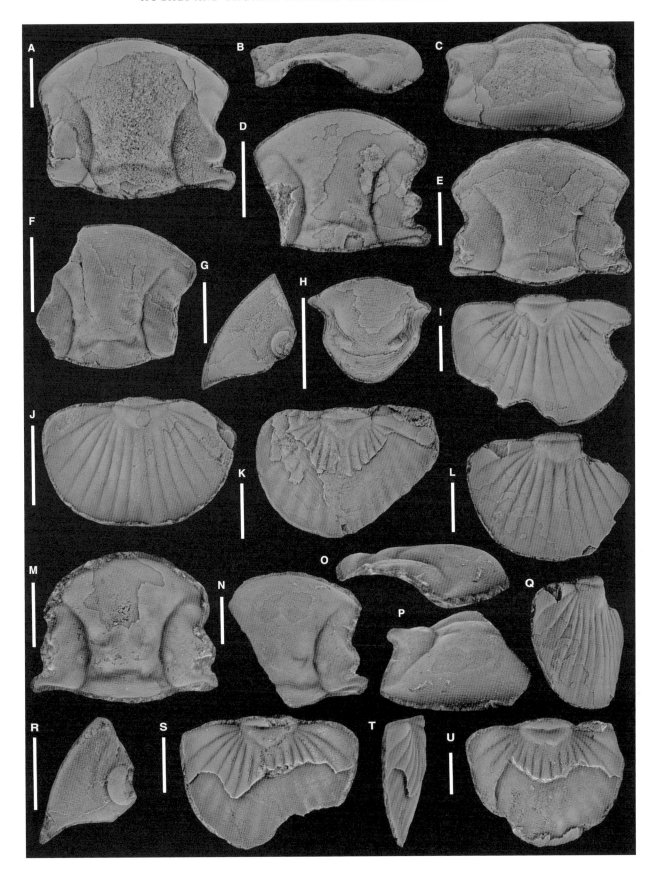

ularly spaced, discontinuous terrace ridges trend transversely across fixigena.

Librigena subtriangular and of low convexity. Epiborder and lateral border furrows equally well impressed, increasingly so anteriorly. Epiborder furrow wider (tr.) than lateral border furrow which sweeps around visual surface, anteriorly converging towards epiborder furrow. Short (exsag.), broad (tr.) genal spine present. Transversely trending terrace ridges.

Pygidium a truncated ellipse, 65% as long (sag.) as wide (tr.) (range = 63–70%; n = 8). Maximum pygidial width at anterior of first pleural rib. Axis triangular with well impressed axial furrows. Axis 47% as long (sag.) as wide (tr.) (range = 42–54%; n = 8), comprising 18% total pygidial length (sag.) (range = 15–21%; n = 8). In lateral view, pygidium smoothly convex; maximum convexity approximately halfway along sagittal length. Median rib roughly twice width (tr.) of lateral pleural ribs at posterior margin. Interpleural furrows clearly impressed, deepest anteriorly. Ribs transversely convex; wider towards margin, which they intersect. Sculpture comprising regularly spaced, transversely trending terrace ridges; mostly continuous across individual pleural ribs and axis. Doublure extending a little over two-thirds total pygidial sagittal length. Ribs and furrows mirror those on the ventral surface, covered in persistent terrace ridges which become increasingly scalloped marginally. Maximum pygidial sagittal length measured 12.4 mm.

Remarks. Jungosulcus ventricosus differs from both *J. emilyae* and *J. anaphalantos* in having: a more convex cranidium and pygidium; axial furrows that maintain their depth of impression throughout course; a shorter (exsag.) anterolateral border; the glabella more expanded anteriorly, so that the cephalon is slightly longer (sag.) relative to its width (tr.); more convex pygidial pleural ribs; maximum pygidial width located at the anterior, rather than the midlength of the first pleural rib; and the median pygidial rib merged with lateral pleural ribs five, six and seven. In addition, *J. emilyae* and *J. anaphalantos* both appear to lack an occipital node, although this is possibly preservational.

Occurrence and distribution. Confined to Valdemar Glückstadt Land (VG1).

Genus MEROPERIX Lane, 1972

Type species. By original designation; *Meroperix ataphrus* Lane, 1972, from the Telychian Samuelsen Høj Formation of Kronpins Christians Land, eastern North Greenland.

Other species. Meroperix aquilonaris Norford, 1981.

Diagnosis. Modified from Lane (1972, p. 343) to take account of data published subsequently by Norford (1981, p. 6): glabella narrowing forward over posterior half; anterior border and preglabellar furrow laterally distinct and very short (exsag.), not present over medial three-fifths of frontal lobe. Pygidium of low convexity, axis about twice as wide (tr.) as long (sag.), occupying about one-fifth pygidial sagittal length; seven pairs of weakly convex pleural ribs; median rib wholly bifid in large holaspides, only the posterior three-fifths bifid in smaller specimens.

Meroperix cf. *ataphrus* Lane, 1972
Figure 11A

Material. MGUH30600 pygidium, from GGU 275038 (VG1).

Remarks. A fragmentary pygidium is assigned to *Meroperix* cf. *ataphrus*. The pygidium differs from pygidia of *M. ataphrus* figured by Lane (1972, pl. 60, figs 4a–c, 5, 6a–b, 10–11) by lacking a faint median furrow continuing from the point of bifurcation of the median rib to the posterior of the axis. The length to width ratio of approximately 2:3 and the low pygidial convexity distinguish the specimen from those assigned to species of *Ekwanoscutellum*.

Genus OPOA Lane, 1972

Type species. By original designation; *Opoa adamsi* Lane, 1972, from the Telychian Samuelsen Høj Formation of Kronpins Christians Land, eastern North Greenland.

Other species. Opoa limatula sp. nov.; *O. magnifica* (Teichert, 1937); *O. ostreata* Lane, 1988; *O. regale* (Fritz, 1964).

Diagnosis. Modified after Lane (1972, p. 340): lateral glabellar muscle impressions adjacent to axial furrow; S1 largest and bordering S2; S3 closer to S2 than to anterior border. Narrow anterior border and weak preglabellar furrow extend along whole frontal lobe, but best developed anterolaterally. Pygidium and pygidial axis both

FIG. 11. A, *Meroperix* cf. *ataphrus* Lane, 1972; MGUH30600 pygidium, GGU 275038 (VGL1) dorsal view. B–O, *Opoa limatula* sp. nov.; B–C, MGUH30601 holotype cranidium; B, dorsal and C, oblique lateral views; D, G, MGUH30602 cranidium; D, lateral and G, dorsal views; E–F, MGUH30603 cranidium; E, dorsal and F, dorsoanterior views; H, MGUH30604 cephalon, dorsal view; I–J, MGUH30605 pygidium, I, dorsal and J, oblique dorsolateral views; K, MGUH30606 pygidium, dorsal view; L–M, MGUH30607 pygidium, L, dorsal and M, lateral views; N, MGUH30608 pygidium, dorsal view; O, MGUH30609 pygidium, dorsal view. All material from GGU 274689 (CPL5). P–R, *Opoa magnifica* (Teichert, 1937), MGUH30610 cranidium, GGU 198842 (WPL2); P, dorsal, Q, oblique lateral and R, dorsoanterior views. Scale bars represent 5 mm (A–O) and 10 mm (P–R).

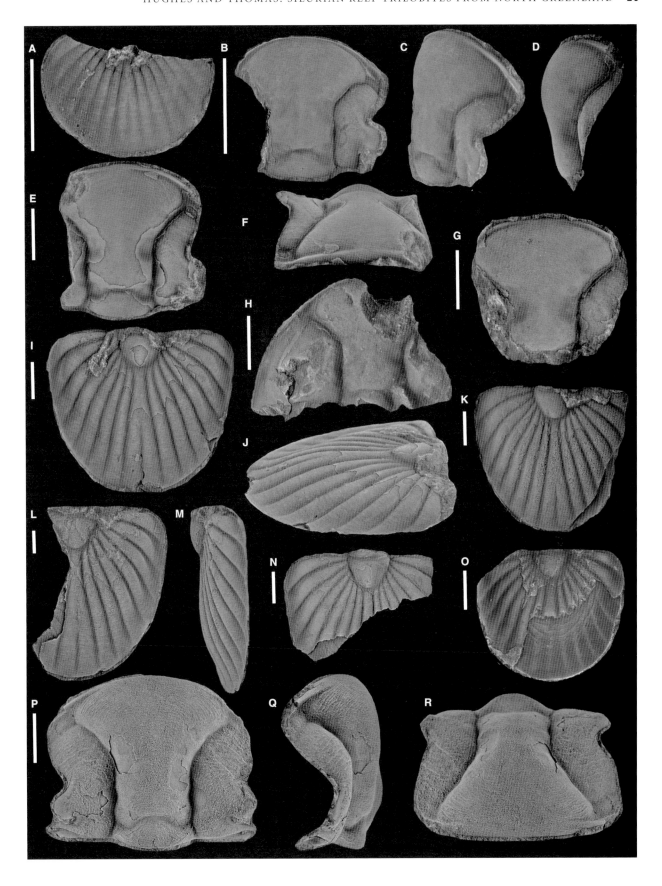

wider than long. Pygidial axis a highly convex blunted triangle comprising one-fifth sagittal length of pygidium. Maximum convexity of pygidium halfway between axis and margins, creating a humpbacked profile.

Remarks. Lane (1972) proposed *Opoa* based on cranidia and pygidia and stressed the distinguishing features as being the arrangement of the glabellar muscle impressions, the presence of an anterior median pit and the distinctive honeycomb sculpture. Lane (1988, p. 96) noted that the sculptural features of *O. adamsi* are superficial and the most important distinguishing characters of the genus are those relating to the axis, particularly the position of the lateral glabellar muscle impressions. As sculpture only has specific importance within this genus, reference to it is removed from the generic diagnosis. Recognition of some other characters from the original diagnosis in other scutelluid taxa here described, for example the anterior median depression, has resulted in their removal also. As well as the named species listed above, Norford (1981, p. 7, pl. 1, figs 10–13) figured several incomplete pygidia and a librigena of *Opoa* sp. from the Telychian – ?lower Wenlock Attawapiskat Formation, Ontario. *Opoa*(?) *trinodosa* Kobayashi and Hamada, 1986, was later transferred to *Borenoria* Holloway and Lane, 2012 (table 2, p. 421).

Opoa limatula sp. nov.
Figure 11B–O

LSID. urn:lsid:zoobank.org:act:026D3357-E467-4419-9B16-638A259D7982

Derivation of name. Latin, *limatulus*, somewhat filed, polished, smoothed, alluding to the lack of the tubercular sculpture found in other species of the genus.

Holotype. MGUH30601 (Fig. 11B–C) cranidium; from GGU 274689 (CPL5).

Figured paratypes. Locality CPL5: MGUH30604 cephalon; MGUH30602–30603 cranidia; MGUH30605–30609 pygidia.

Diagnosis. Anterior border and preglabellar furrow very weak medially. Pygidium with median rib bifid for most of its posterior half. Pygidial margins smoothly curved. Sculpture of terrace ridges.

Description. Cephalon semicircular. Cranidium of roughly equal width posteriorly between fulcral sockets as at anteriorly between anterolateral margins; width across β–β and δ–δ approximately equal. Width (tr.) across γ–γ about equal to cranidial sagittal length. Cranidium just over four-fifths as long (sag.) as wide (tr.) across palpebral lobes at δ–δ, roughly 33% as high as long

(sag.). Highest point of cranidium in anterior part of glabella. Where preservation permits, subdued median carina apparent (Fig. 11B). Cranidial sagittal length 88% width (tr.) across palpebral lobes (range = 77–99%; n = 5). Glabella parallel-sided for its posterior 49% (range = 45–50%; n = 4); between S1 and S2, axial furrows suddenly change course and diverge rapidly forward at 44–47 degrees (n = 4) to an exsagittal line, reaching maximum width (tr.) where intersects anterolateral border furrow; there, glabella 2.1 times wider (tr.) than at its narrowest point (range = 2–2.2 times; n = 5). Occipital ring highest posteriorly and longest sagittally, where 38% as long (sag.) as wide (tr.) (range = 37–40%; n = 4). Narrow (sag.), deeply impressed occipital furrow. Abaxially, occipital ring shortens to accommodate large occipital impression. Narrow (sag., exsag.) anterior border and preglabellar furrow extend along whole frontal lobe; both very weak medially. Abaxially, preglabellar furrow becomes deeper and wider. S1 most impressed of lateral glabellar muscle impressions; reniform, with median node. S2 in very close proximity to S1; subsquare. S3 transversely directed ellipse, further from S2 than S2 is from S1. S2 and S3 of comparable size. Whereas S1 is connected to axial furrow, S2 and S3 are distinct from it (Fig. 11D, G). Sculpture of persistent terrace ridges; following curvature of anterior margin, trending transversely, most pronounced along anterior of glabella, fading posterior of S2. Largest cranidial sagittal length measured 12.1 mm.

Fixigena with palpebral lobe extending from anterior of occipital muscle impression to posterior of S1; lunette from opposite anterior of occipital muscle impression to posterior of S1. Eye ridge trending towards axial furrow near S3. Sculpture of terrace ridges trending transversely.

Librigena with narrow epiborder furrow most pronounced anteriorly. Epiborder furrow shallow, continuing through to short genal spine. Lateral border furrow deeply impressed, gently converging towards epiborder furrow anteriorly.

Pygidium an anteriorly truncated ellipse with smoothly curved margins, 86% as long (sag.) as wide (tr.) (range = 80–96%; n = 6). Axis faintly trilobed, with well impressed axial furrows. Axis comprising 21% entire sagittal length of pygidium (range = 20–24%; n = 7). Anterior (maximum) width of axis 71% of sagittal length (range = 58–77%; n = 6). Pleural region slopes upwards (tr. and sag.) abaxially, until a point approximately halfway from the margins, from which it slopes downwards (tr. and sag.) abaxially. At posterior margin, bifid median rib is twice width of pleural ribs at margins. Pleural ribs smoothly transversely convex. Interpleural furrows and pleural ribs widen towards margin where furrows are fainter. Sculpture of transversely trending, short terrace ridges covering pleural region and axis. Doublure extending halfway to axis, with well impressed ribs and furrows mirroring dorsal surface. Well developed terrace ridges become increasingly scalloped towards margins. Maximum pygidial sagittal length measured 24.5 mm.

Remarks. The type species of *Opoa*, *O. adamsi*, is a very distinctive scutelluid, particularly because of its honeycomb sculpture and scalloped pygidial outline. These features are superficial, however. Similarities in shape and arrangement of the glabellar muscle impressions, and the inflation and form of the pygidial axis, confirm the generic assignment of *O. limatula*. The shape of the glabella and the relative proportions of both the pygidium and pygidial axis are synapomorphies also. As well as lacking the distinctive sculpture and pygidial outline of *O. adamsi*, *O. limatula* differs in the absence of an anterior medial pit.

Occurrence and distribution. Most abundant in Central Peary Land (CPL5). Also present in Kronprins Christian Land (KCL1).

Opoa magnifica (Teichert, 1937)
Figures 11P–R, 12, 13A–C

v*. 1937 *Scutellum magnificum* Teichert, p. 148–149, pl. 21, figs 5–6.

Holotype. MGUH4375 (Fig. 12L) pygidium; from St George Fjord.

Diagnosis. Pygidium roughly as wide as long, of low convexity, with long (exsag.), well developed anterolateral concave zone; pygidial axis less convex than in other species of *Opoa*; median rib a little wider (tr.) distally than lateral ribs, median rib either non-bifid or with a weak sagittal furrow.

Description. Cephalon semicircular, almost twice as wide (tr.) as long (sag.); highest point in anterior part of glabella. Cranidium of approximately equal width at β–β and ω–ω, wider across β–β than δ–δ. Width (tr.) at δ–δ a little greater than sagittal length of cranidium (89% as long as wide; range = 81–97%; n = 15); width (tr.) across γ–γ slightly less than cranidial sagittal length. Posterior half of glabella parallel-sided, anterior of S1 expanding rapidly forwards at 35–40 degrees to sagittal line (n = 3). Anterior course of axial furrow smoothly curved, intercepting anterolateral border furrow where glabella is just over twice posterior width (tr.) (2.2 times wider than at its narrowest part (range = 2–2.5 times; n = 10)). Facial suture positioned closer to axial furrow, relative to sagittal line, at α than at γ. Occipital ring about 40% as long (sag.) as wide (tr.) (range = 36–44%; n = 12); slopes forwards; just over twice as long sagittally as exsagittally. Occipital furrow strongly impressed; occipital impression large. Anterior border and preglabellar furrow variably developed medially. Glabellar muscle impression S1 largest, most impressed, ovoid, with large, low, central swelling. S2 and S3 transversely elongate, of comparable size; S3 positioned further from S2 than S2 is from S1. S1 in contact with axial furrow, S2 and S3 close to axial furrow, but not meeting it. Sculpture comprises transversely trending terrace ridges following anterior margin convexity; most distinctive in L3. Maximum cranidial sagittal length measured 32.7 mm.

Palpebral lobe located between occipital furrow and S1. Lunettes oval, located adjacent to L1. Eye ridge intersecting axial furrow near S3. Anterior portion of fixigena with transversely trending terrace ridges; posterior to eye ridge with pronounced elongate pits on internal moulds (Fig. 12C) and irregular obliquely trending terrace ridges on external surface (Fig. 11P).

Librigena with prominent lateral border, narrowing (tr.) posteriorly. Epiborder furrow well impressed, wide (tr.); lateral border furrow sharply defined, parallel to lateral border over anterior half of librigena; over posterior half, diverging from outer portion of lateral border. Visual surface comprises roughly one-third of entire sagittal length of cephalon. Librigena terminating in slender genal spine, comprising roughly 20% total cephalic sagittal length; terrace ridges converging towards tip of spine. External moulds show deeply impressed genal terrace ridge whorl on doublure (Fig. 12K).

Rostral plate with slightly curved connective sutures diverging at 40–50 degrees to sagittal line. Sculpture of regularly spaced, continuous terrace ridges trending transversely.

Hypostome typical for the family; subtriangular, with posterior margin terminating in blunted point. Sagittal length roughly 80% of width (tr.) across anterior wings; short posterior lobe of middle body; prominent maculae, with midpoint located about halfway along hypostome length; posterior border longer (sag.) than lateral border is wide (tr.). Prominent terrace ridges cover entire hypostome, medially trending transversely, abaxially following line of borders.

Pygidium an anteriorly truncated ellipse, 93% as long as wide (range = 79–111%; n = 19); when plotted, the pygidial transverse width = 0.88 pygidial sagittal length +2.55 (see Remarks for discussion, and Fig. 7A); with maximum width close to anterior margin. Pygidial axis semi-elliptical in outline and flattened; 91% as long as wide (range = 77–109%; n = 46); when plotted, the pygidial axis transverse width = 1.12 pygidial axis sagittal length −0.11 (see Remarks for discussion, and Fig. 7B), and comprising 23% pygidial sagittal length (range = 19–30%; n = 28); when plotted, the pygidial axis sagittal length = 0.20 pygidial sagittal length +0.30 (see Remarks for *E. agmen* for discussion, and Fig. 7C); faintly trilobate; well impressed axial furrows. Pleural ribs with low transverse convexity, sigmoidal, widen towards margins, which they intersect. Interpleural furrows narrow and distinct proximally, becoming slightly weaker and wider marginally. From axis, median rib narrowing for about one-third its length, after which it widens or becomes subparallel-sided distally. Sculpture comprising irregularly spaced tubercles accompanied by weak, discontinuous terrace ridges present only near posterior and lateral margins (Fig. 12Q–R). Pygidial doublure with prominent terrace ridges; progressively scalloped towards margins. Largest pygidium 59.1 mm long.

Remarks. Specimens of *O. magnifica* display the following intraspecific variation: the extent of the anterior border and preglabellar furrow is very variable – they may be present for all (Fig. 12F), or as little as half of the anterior cranidial width (tr.) (Fig. 12G) – and not correlated with specimen size. When a particularly well developed anterior border is present, a median anterior depression is seen (Fig. 12A–B). The median rib on the pygidium is most commonly non-bifid, but a weak sagittal furrow may be present distally (Fig. 12P–Q). On the pygidial doublure, either a faint sagittal ridge is present or the terrace ridges are distorted medially (Fig. 13A). Pygidia decrease in convexity with increasing size (compare Figs 12R, 13C), as do

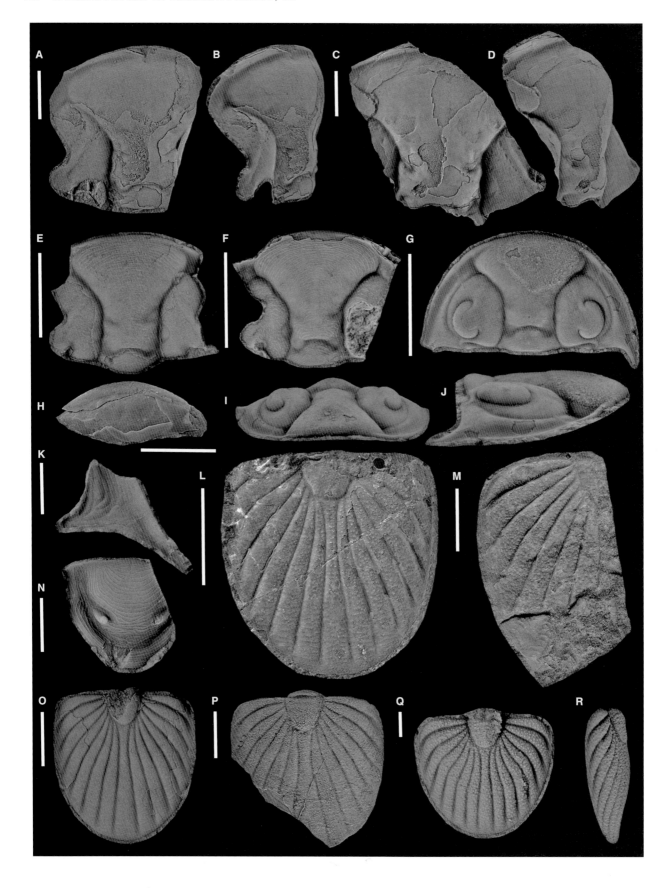

cranidia; pygidia become progressively proportionally longer with growth – the smallest pygidia are a little wider than long, while larger pygidia are a little longer than wide (pygidial sagittal lengths average 83% of pygidial transverse width in the five smallest specimens, compared with 105% in the five largest specimens).

Teichert (1937, p. 148) proposed the species *Scutellum magnificum* based on a single pygidium (pl. 21, fig. 6, MGUH4375), which is thus the holotype by monotypy (Fig. 12L), from the Offley Island Formation (Telychian) from St George Fjord, North Greenland. This pygidium is considered conspecific with material here described from Western Peary Land: pygidial proportions, axial morphology and the faintly bifid median rib are identical. The holotype is convex and covered with tubercles, as would be expected in a specimen of the species at this size. Teichert assigned another pygidium (pl. 21, fig. 5, MGUH4374) from Kûk, mouth of Thomsen River, Duke of York Bay, northern Southampton Island to *S. magnificum* with some uncertainty. We have studied the pygidium (Fig. 12M), and although it is fragmentary, its proportions and sculpture confirm its assignment to *O. magnifica*.

The cephalon of *O. magnifica* is closest to *O. limatula*, particularly in glabellar morphology. The most notable cephalic differences are that the anterior border and preglabellar furrow are weaker medially in *O. magnifica*, and the anterolateral concave zone is longer (exsag.). Additionally, the glabella of *O. magnifica* is less convex than that of *O. limatula*. Smaller pygidia of *O. magnifica* are remarkably similar to those of *O. ostreata* in outline, proportions and axial morphology. The most obvious differences are the lack of bifurcation of the median rib in *O. magnifica*, and the relative widths of the median rib at the margins: that of *O. ostreata* is twice as wide as that of *O. magnifica*. The sculpture of smaller specimens of *O. magnifica* shares the following similarities with smaller specimens of *O. ostreata*: both have terrace ridges covering the cephalon which are shortest on the librigena, and the pygidia bear tubercles on the pleural regions and short terrace ridges on the margins. Pygidia of *O. magnifica* may be similar in size to those of *Ekwanoscutellum* species. The pygidium of *O. magnifica* resembles that of *Ekwanoscutellum* in terms of its relative proportions sufficiently that they can be confused with one another, even

though they are not congeneric. However, the lower convexity of the pygidium of *O. magnifica*, combined with a posteriorly narrower median rib, and significant changes in sculpture with ontogeny distinguish it. The resulting straight-line equations for pygidial axial width against pygidial axial length of *O. magnifica* and *E. agmen* are very similar for the two species (Fig. 7B). Conversely, the straight-line equations for pygidial width against pygidial length show the width of the pygidium to increase at a slower rate with increasing length for *O. magnifica* than for *E. agmen* (Fig. 7A).

Occurrence and distribution. Equally common in Western Peary Land (WPL1–2) and Wulff Land (WL). Less common in South Nares Land (SNL1–2).

Opoa cf. *magnifica* (Teichert, 1937)
Figure 13D–E

Remarks. Two cephala resemble those of *O. magnifica*, particularly in the narrow posterior part of the glabella, pronounced anterior border, and low convexity glabella. They differ, however, in the extremely narrow posterior half of the glabella, very pronounced anterior border extending across the entire anterior margin, and very pronounced lateral border. There also similarities with *O. limatula* (Fig. 11B–O), particularly the glabellar furrows, which exhibit an angular change in course at the point of lateral glabellar expansion. However, the glabella of *O. limatula* is much wider (tr.) posteriorly, and more forwardly expanded, leaving only a weak anterior border medially. *Opoa limatula* is also more convex.

Occurrence and distribution. Central Peary Land (CP3).

Opoa sp.
Figure 13F–I

Remarks. Two fragmentary pygidia with sagittal lengths over 20 mm are assigned to *Opoa* because of their proportions (measurable pygidium is 0.9 times as long (sag.) as wide (tr.)), the inflated form of the axis and the

FIG. 12. *Opoa magnifica* (Teichert, 1937). A–B, MGUH30611 cranidium, GGU 298504 (WL); A, dorsal and B, oblique lateral views. C–D, MGUH30612 cranidium, GGU 298504 (WL); C, dorsal and D, oblique lateral views. E, MGUH30613 cranidium, GGU 301326 (WPL1) dorsal view. F, MGUH30614 cranidium, GGU 301318 (WPL1) dorsal view. G, I–J, MGUH30615 cephalon, GGU 298506 (WL); G, dorsal, I, dorsoanterior and J, lateral views. H, MGUH30616 rostral plate, GGU 301318 (WPL1) ventral view. K, MGUH30617 librigena doublure, GGU 301318 (WPL1) dorsal view. L, MGUH4375 holotype pygidium, St George Fjord, dorsal view. M, MGUH4374 pygidium, Duke of York Bay, northern Southampton Island, dorsal view. N, MGUH30618 hypostome, GGU 298506 (WL) ventral view. O, MGUH30619 pygidium, GGU 301318 (WPL1) dorsal view. P, MGUH30620 pygidium, GGU 198834 (WPL2) dorsal view. Q–R, MGUH30621 pygidium, GGU 301326 (WPL1); Q, dorsal and R, lateral views. Scale bars represent 10 mm (A–F, L–M, O–P), 5 mm (G–K, N) and 1 mm (Q–R).

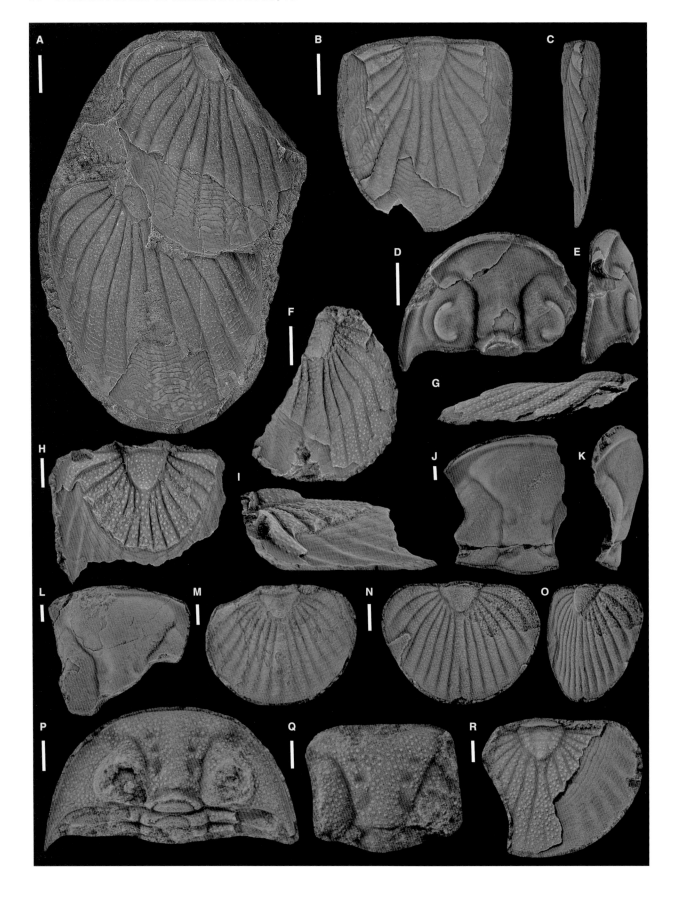

humpbacked profile. Synapomorphies with *O. adamsi* Lane, 1972 include the extent of the axis, which comprises one-quarter entire sagittal length of the pygidium, the bifid median rib, and the extent and morphology of the doublure. Differences from *O. adamsi* include the smoothly curved, not scalloped, margin, and the less slender axis, which is roughly as long (sag.) as wide (tr.) at its most anterior point, not longer than wide as in *O. adamsi*. The complete covering of pronounced, irregularly spaced tubercles of *O. sp.* are reminiscent of those in *O. adamsi* (see Lane 1972, pl. 59), *O. regale* (Fritz, 1964; see Ludvigsen 1979, fig. 46c) and *Opoa* sp. (of Norford, 1981, see pl. 1, figs 10–13; pl. 5, figs 14, 16–18).

Occurrence and distribution. Kronprins Christian Land (KC1).

Scutelluid gen. indet. 1
Figure 13J–O

Remarks. Two cranidia differ from those of other scutelluids in the positioning of the fixigenal equivalent of the librigenal inner furrow, which is clearly separated from the long (exsag.) anterolateral concave zone. The cranidia are closest to *Ekwanoscutellum*, particularly with respect to the position and morphology of the lateral glabellar muscle impressions: S1 is about the same size as the well impressed occipital impression, and S2 is a transversely directed oval shape, about one-quarter the size of S1. S2 is positioned directly anterior of S1, and is not connected to the axial furrow. S3 is situated much further anteriorly of S2, than S2 is of S1, is similar in size to S2 and is also close to the axial furrow. The axial furrows, which exhibit an angular change in course at the point of glabellar expansion, are also similar to those of *Ekwanoscutellum*. The prominent nature of the anterior border, with an anterior median depression, and the long anterolateral concave zone are more similar to *Opoa*, however.

Two pygidia measuring 4.7 and 5.3 mm sagittal length, and both 72% as long (sag.) as wide (tr.), are potentially conspecific with the cranidia on the basis of their co-occurrence. The axis is a blunted triangle comprising roughly one-quarter the pygidial sagittal length, and the median rib is bifid. Differences in sculpture between the

two specimens are likely ontogenetic: the larger specimen (Fig. 13N–O) has a predominantly tuberculate sculpture, except for faint terrace ridges near edges, and the smaller specimen (Fig. 13M) has well spaced terrace rides transversely trending across the whole pygidium. The pygidia have significantly different proportions from those of *Ekwanoscutellum* and are more similar in this respect to *Meroperix*, although the position of greatest pygidial width (tr.) is positioned slightly more anteriorly than in *Meroperix*.

Occurrence and distribution. Kronprins Christian Land (KCL2).

Scutelluid gen. indet. 2
Figure 13P–R

Remarks. A cranidium and cephalon measuring 4.1 and 4.6 mm in sagittal length, respectively, and a pygidium measuring 6.1 mm in sagittal length, have low convexities and are covered in coarse, dense tubercles (pygidial tubercles bear central pores). They do not readily fit into any known scutelluid genus. The small size of the sclerites, cephalic proportions (41% as long (sag.) as wide (tr.) across its widest point) and almost straight anterior cephalic margin suggests that these are not fully grown holaspids. The shape and positions of the lateral glabellar muscle impressions are closest to *Jungosulcus* species, but they are equidistant from the sagittal line; S1 is kidney-shaped with a median node and is connected to the axial furrow; S2 and S3 more oval in form and directed transversely, with S2 slightly wider (exsag.) and S3 more transversely slit-like. The pygidium is about three-quarters as long (sag.) as wide (tr.). The triangular axis is 70% as long (sag.) as wide (tr.), comprising about one-quarter of the total pygidial sagittal length, and the median rib is bifid.

Occurrence and distribution. Western Peary Land (WPL1).

Subfamily BUMASTINAE Raymond, 1916

Remarks. See Remarks under Family Illaenidae and Family Scutelluidae.

FIG. 13. A–C, *Opoa magnifica* (Teichert, 1937); A, MGUH30622 two stacked pygidia, GGU 301318 (WPL1) dorsal view; B–C, MGUH30623 pygidium, GGU 301326 (WPL1); B, dorsal and C, lateral views. D–E, *O.* cf. *magnifica* (Teichert, 1937), MGUH30624 cephalon, GGU 198225 (CPL3); D, dorsal, and E, lateral views. F–I, *Opoa* sp.; F–G, MGUH30625 pygidium; F, dorsal and G, lateral views; H–I, MGUH30626 pygidium; H, dorsal and I, lateral views; both pygidia from GGU 275015 (KCL1). J–O, Scutelluid gen. indet. 1; J–K, MGUH30627 cranidium, GGU 225793 (KCL2); J, dorsal and K, lateral views; L, MGUH30628 cranidium, GGU 225793 (KCL2) dorsal view; M, MGUH30629 pygidium, GGU 225793 (KCL2) dorsal view; N–O, MGUH30630 pygidium, GGU 225793 (KCL2); N, dorsal and O, oblique lateral views. P–R, Scutelluid gen. indet. 2; P, MGUH30631 cephalon, GGU 301318 (WPL1) dorsal view; Q, MGUH30632 cranidium, GGU 301318 (WPL1) dorsal view; R, MGUH30633 pygidium, GGU 301318 (WPL1) dorsal view. Scale bars represent 10 mm (A–C), 5 mm (D–I) and 1 mm (J–R).

Genus CYBANTYX Lane and Thomas *in* Thomas, 1978

Type species. By original designation; *Cybantyx anaglyptos* Lane and Thomas *in* Thomas, 1978, from the upper Homerian Much Wenlock Limestone Formation, Dudley, West Midlands, UK.

Other species. Cybantyx cuniculus (Hall, 1867); *C. harrisi* (Weller, 1907); *C. holmi* (Lindström, 1885); *C. insignis* (Hall, 1864); *C. nebulosus* sp. nov.; *C. niagarensis* (Whitfield, 1880).

Diagnosis. Emended from Lane and Thomas *in* Thomas (1978, p. 18) to include their description of lateral glabellar muscle impressions in the type species: effaced scutelluid with pronounced anterior border furrow and short, sharply upturned anterior border. S1 and occipital impression close together; S1 largest, but only slightly larger than occipital impression; S2 smallest; S2 and S3 transversely elongate. Axial furrow impressed posterior of well impressed omphalus. Omphalus with median granule. Small anterolateral internal pit. Rounded genal angle. Pygidium with sagittal carina posteriorly.

Remarks. Ludvigsen and Tripp (1990) erected the genus *Paracybantyx* to distinguish their species, *P. asulcatus* from *Cybantyx*, due to its lack of an anterior border and border furrow. *Paracybantyx* has since been considered a junior subjective synonym of *Failleana* (Curtis and Lane 1997). Adrain *et al.* (1995, p. 729) considered the distinguishing morphological character of *Paracybantyx* to be a rostral plate lacking a posterior lobe and dorsal flange, and on this basis assigned their Alaskan species *P. occidentalis* Adrain, Chatterton and Blodgett, 1995, which has a rostral plate displaying a forwardly convex median embayment, to the genus. However, as Adrain *et al.* noted, the rostral plate of the type species of *Paracybantyx, P. asulcatus* Ludvigsen and Tripp, 1990, is unknown, and the lack of a posterior lobe and dorsal flange in that species is presumed from the morphology of other Wenlock–Ludlow northern Laurentian illaenimorphs. More recently, Chatterton and Ludvigsen (2004) described two new species of *Failleana* from Anticosti Island, Québec, Canada: *F. magnifica* and *F. wangi*, both of which exhibit a rostral plate with a forwardly convex median embayment, and this is the condition in *C. nebulosus* sp. nov. (Figs 14A–O, 15A–J) also. Adrain *et al.* (1995) considered this form of hypostomal attachment to be a derived character state, potentially defining a northern Laurentian

clade and questioned the use of the anterior border as a diagnostic character. Pending further knowledge of the distribution of character states – including lateral glabellar muscle impressions and rostral plate morphology – in the taxa concerned, we retain the restricted definition of *Cybantyx* as in our emended diagnosis.

Cybantyx nebulosus sp. nov.
Figures 14, 15A–J

LSID. urn:lsid:zoobank.org:act:B7A44B7F-A714-4316-862A-39FC74E42770

Derivation of name. Latin, *nebulosus*, misty, cloudy, dark, indefinite, alluding to the difficulty of establishing the affinities of the species and genus.

Holotype. MGUH30635 (Fig. 14D–E) cephalon; from GGU 274689 (CPL5).

Figured paratypes. Locality CPL5: MGUH30634 articulated specimen; MGUH30642 articulated cephalon and thorax; MGUH30636, 30638–30639 cephala; MGUH30637, MGUH30640 cranidium; MGUH30641 librigena; MGUH30643, 30645 rostral plates; MGUH30644, 30646–30649 pygidia; MGUH30650 pygidium with articulated thorax.

Diagnosis. Cranidial length (sag.) roughly equal to width (tr.) across palpebral lobes. Lunette opposite midpoint of palpebral lobe. Small reniform impression situated between the occipital impression and S1. Rostral plate with a forwardly convex median embayment. Pygidium wider (tr.) than long, and with holcos well developed.

Description. Cephala and cranidia ranging in sagittal length from 2.9 to 15 mm. Cephalon an anteriorly elongated semicircle; about 80% as high as long (sag.), with maximum height in palpebral view opposite posterior of palpebral lobe. Cephalon 60% as long (sag.) as wide (tr.) in palpebral view (range = 53–69%; n = 7). Sagittal carina on internal moulds, extending from occipital node and fading out before anterior margin. Anterior margin upturned forming thin, pronounced, rim-like anterior border, border furrow even narrower (sag.). Cranidium 98% as long (sag.) as wide (tr.) across palpebral lobes (range = 87–107%; n = 22). Glabella narrower posteriorly than width of cranidium at δ–δ. Axial furrow anteriorly curves adaxially, intersecting lunette; there positioned 76% of width apart (tr.) at posterior margin (range = 68–81%; n = 17). Lunette an exsagit-

FIG. 14. *Cybantyx nebulosus* sp. nov. A–C, MGUH30634 articulated specimen; A, dorsal view of whole specimen; B, dorsal view of cephalon; and C, oblique view of whole specimen. D–E, MGUH30635 holotype cephalon; D, dorsal and E, lateral views. F, MGUH30636 cephalon, dorsal view (not blackened). G, MGUH30637 cranidium, dorsal view. H–I, K, MGUH30638 cephalon; H, dorsal, I, lateral and K, anterior views. J, MGUH30639 cephalon, oblique lateral view. L, MGUH30640 cranidium, dorsal view. M, MGUH30641 librigena, dorsal view. N–O, MGUH30642 articulated cephalon and thorax; N, dorsal and O, lateral views. All material from GGU 274689 (CPL5). Scale bars represent 10 mm (A–L, N–O) and 1 mm (M).

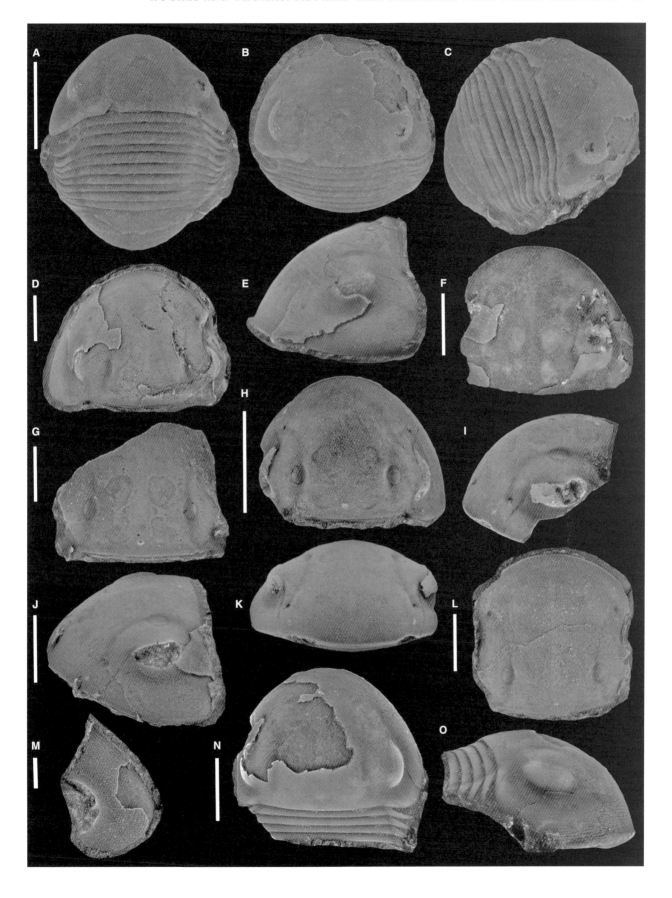

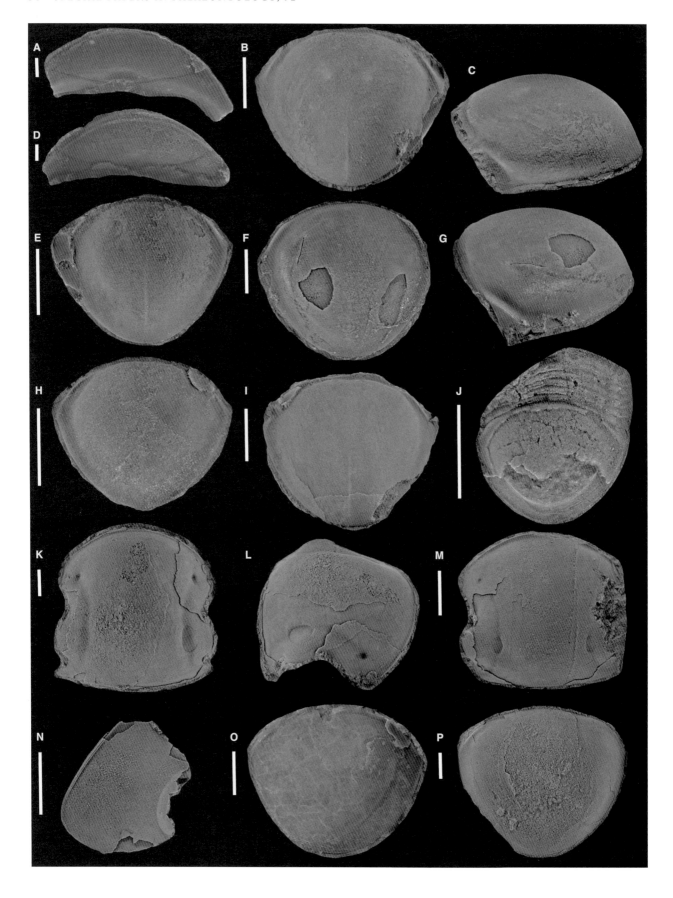

tally directed oval, very well impressed on internal mould, less so on external. Anterior of lunette, axial furrow curves abaxially, becoming gradually closer to facial suture and terminating at omphalus, where positioned 154% of width apart (tr.) at lunettes (range = 140–169%; n = 15). On internal moulds, axial furrow distinct throughout course, best impressed between lunette and omphalus. Axial furrows largely effaced on external moulds. Omphalus and median granule present, clear on internal mould, much fainter externally. Anterolateral internal pit smaller than omphalus, only visible on internal mould; slightly closer to anterior margin than to omphalus and placed a little closer to sagittal line. Occipital node pronounced on interior mould, much fainter externally.

Lateral glabellar muscle impressions variably expressed within population; only visible on internal mould and rarely clear. Occipital impression kidney-shaped, larger than lunette, and extending from opposite occipital node to posterior of lunette. S1 extending from opposite midpoint of lunette to anterior margin of palpebral lobe. Anteriorly, muscle impressions situated progressively further from the sagittal line (Fig. 14G).

External surface with laterally continuous, non-anastomosing terrace ridges, evenly spaced, running parallel to anterior margin, strongest at anterior margin, not extending farther back than front of palpebral lobe. Palpebral lobe comprising roughly one-quarter entire sagittal length, gently rounded, following curvature of cranidium with same convexity.

Librigena evenly convex, with gently rounded genal angle. Thin border present anteriorly. Finely pitted throughout, terrace ridges present near margins. Terrace ridges strongest and most evenly spaced anteriorly; weaker and increasingly shorter around genal angle.

Rostral plate with low convexity; maximum convexity where upturned to meet anterior margin. Connective sutures very slightly curved, strongly diverging forwards. Posteromedial depression on internal mould (Fig. 15A). Cuticle with regularly spaced, strong, continuous terrace ridges running parallel to anterior margin.

Pygidia ranging in sagittal length from 2.1 to 17.1 mm. Pygidium about 70% as high as long (sag.), with maximum height approximately halfway along sagittal length (extended view); rounded triangular in shape. Pygidium averaging 82% as long as wide (range = 73–89%; n = 41); when plotted, the pygidial transverse width = 1.15 pygidial sagittal length +0.67 (see Remarks for discussion, and Fig. 16). Sagittal carina variably preserved; most commonly just over posterior half of pygidium, but may be absent or present for most of sagittal length. Holcos best developed anterolaterally. Abaxial end of articulating facet located opposite roughly one-third pygidial length (sag.) from anterior. Pair of swellings on internal mould, representing mus-cle impressions, rarely preserved (Fig. 15B); located opposite abaxial end of articulating facet. Cuticle obscures sagittal carina, with irregular, discontinuous terrace ridges trending transversely. Terrace ridges strongest at margins. Doublure approximately 25% sagittal length of pygidium and uniform width (Fig. 15J).

Ontogeny. The pygidium of *C. nebulosus* becomes progressively longer with growth; pygidial sagittal lengths average 78% of pygidial transverse width in the five smallest specimens, compared with 85% in the five largest specimens.

Remarks. The size, shape and positions of the lateral glabellar muscle impressions of *C. nebulosus* are strikingly similar to those of the type species, *C. anaglyptos.* The small reniform impression situated between the occipital impression and S1 of *C. nebulosus* (Fig. 14G) is intermediate between the separated occipital impression and S1 of *C. nebulosus* and the conjoined form of *Bumastus* Murchison, 1839. Other synapomorphies include a pronounced anterior border, well developed omphalus and sagittal pygidial carina.

The main differences between *C. nebulosus* and *C. anaglyptos* are as follows: the cephalon of *C. nebulosus* is proportionally slightly longer; the midpoint of the lunette is positioned opposite the midpoint of the palpebral lobe in *C. nebulosus*, and is more anteriorly placed in *C. anaglyptos*; the rostral plate of *C. nebulosus* bears a forwardly convex median embayment, whereas the rostral plate of *C. anaglyptos* bears a broad-based rounded posteromedian projection; the pygidium of *C. nebulosus* is 82% as long as wide, whereas the pygidium of *C. anaglyptos* measures as long (sag.) as it is wide (tr.).

The cephalic morphology of *C. nebulosus* is most comparable to *P. occidentalis* Adrain, Chatterton and Blodgett, 1995, the main distinguishing features being: the cranidium of *C. nebulosus* is roughly as long (sag.) as wide (tr.) across palpebral lobes, in *P. occidentalis,* it is only 84% as long as wide; the lunettes of *C. nebulosus* are less elongated and situated a little farther forward.

The pygidium of *C. nebulosus* is typically illaenimorph and shows strong similarities to a range of species, particularly to *P. occidentalis*, which has a similar length-to-width ratio. The main differences from *P. occidentalis* are that the abaxial end of the pygidial articulating facet of *C. nebulosus* is positioned further posteriorly, there is a

---

FIG. 15. A–J, *Cybantyx nebulosus* sp. nov.; A, MGUH30643 rostral plate, ventral view; B–C, MGUH30644 pygidium; B, dorsal and C, lateral views; D, MGUH30645 rostral plate, ventral view; E, MGUH30646 pygidium, dorsal view; F–G, MGUH30647 pygidium; F, dorsal and G, lateral views; H, MGUH30648 pygidium, dorsal view; I, MGUH30649 pygidium, dorsal view; J, MGUH30650 pygidium with articulated thorax, dorsal view. All material from GGU 274689 (CPL5). K–P, *Cybantyx* sp.; K–L, MGUH30651 cranidium, GGU 301319 (WPL1); K, dorsal and L, oblique lateral views; M, MGUH30652 cranidium, GGU 301319 (WPL1), dorsal view; N, MGUH30653 librigena, GGU 301318 (WPL1) dorsal view; O, MGUH30654 pygidium, GGU 301318 (WPL1) dorsal view (not blackened); P, MGUH30655 pygidium, GGU 301319 (WPL1) dorsal view. Scale bars represent 1 mm (A, D), 10 mm (B–C, E–I, K–P) and 5 mm (J).

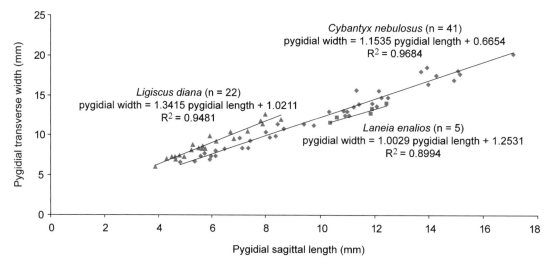

FIG. 16. Major axis regression of pygidial transverse widths against pygidial sagittal lengths for *Cybantyx nebulosus*, *Laneia enalios* and *Ligiscus diana*.

wider (tr.) holcos, and *P. occidentalis* lacks a sagittal carina (possibly preservational).

As with the non-effaced scutelluids, reduced major axis regression was employed to investigate size-related variation and potential for quantitative morphological comparisons of effaced species (Fig. 16). Pygidial width was plotted against pygidial length for taxa with a reasonable sample size: *C. nebulosus* and *Ligiscus diana* sp. nov. *Laneia enalios* sp. nov. is included also, although only five data points are available, as the pygidia are superficially very similar to those of *C. nebulosus*. Different straight-line equations initially suggest that there are differences from *L. enalios*, although more data are required to confirm this. Pygidial width increases with length at a slower rate in *C. nebulosus* than for *L. diana*.

Occurrence and distribution. Found throughout Central Peary Land (CPL1–5).

Cybantyx sp.
Figure 15K–P

Remarks. The cranidia and pygidia of *C.* sp. are both larger than the numerous cranidia and pygidia of *C. nebulosus* described. The cranidia, measuring 17.8 and 24.7 mm in sagittal length, are significantly less convex than those of *C. nebulosus*. Additionally, the axial furrows are better impressed and begin to diverge forwards more anteriorly than in *C. nebulosus*, exhibiting a more sudden change in course; the glabella is slightly less anteriorly expanded; the lunettes are longer (exsag.), and the holcos is less well developed. Pygidia range in sagittal length from 17.5 to 25.4 mm, and are less convex than those of *C. nebulosus*.

Detailed comparison of *C.* sp. with other species of *Cybantyx* is not appropriate given the limited material

available, and the high variability of characters such as the sagittal carina and occipital node in at least some populations, as demonstrated by *C. nebulosus*. The material is therefore left under open nomenclature.

Occurrence and distribution. Most common in Western Peary Land (WPL1–2), also present in South Nares Land (SNL2).

Genus LANEIA gen. nov.

LSID. urn:lsid:zoobank.org:act:EF466B9F-0D6B-48AA-A40F-26AB9AD60381

Derivation of name. For Dr P. D. Lane, studier of smooth trilobites, who collected much of this material; constructed in the style of *Struveria* and *Briania*; gender feminine.

Type species. *Laneia enalios* sp. nov. from the Telychian Samuelsen Høj Formation, Valdemar Glückstadt Land, North Greenland.

Diagnosis. Effaced scutelluid with cephalon of lower convexity than pygidium, with pronounced, rim-like anterior border, and even narrower border furrow. Short, thorn-like genal spine. Eye comprising roughly one-quarter total sagittal cephalic length and placed its own length from posterior of cephalon. Pygidium 90% as long (sag.) as wide (tr.), holcos only defined anterolaterally.

Remarks. The combination of a low convexity cephalon with short genal spines – which have not been reduced to a rounded form as in most illaenimorph taxa – and a pronounced anterior border, distinguish *Laneia* from other illaenimorphs. The eye is relatively short and placed more anteriorly than in most illaenimorph taxa. The cephalon is both less effaced and less convex than the pygidium.

Laneia enalios sp. nov.
Figure 17

LSID. urn:lsid:zoobank.org:act:A977758C–F6D0-4331–A487–1E C45F9D324C

Derivation of name. Greek, *enalios*, in, on, of the sea.

Holotype. MGUH30656 (Fig. 17A–D) cephalon; from GGU 275038 (VGL1).

Figured paratypes. Locality VGL1: MGUH30657–30659 cephala; MGUH30660–30663 pygidia.

Diagnosis. As for genus.

Description. Cephala ranging in sagittal length from 9.8 to 10.5 mm. Cephalon with smoothly rounded anterior margin, 63% as high as long (sag.), with maximum height opposite anterior of palpebral lobe in palpebral view. Cephalon 76% as long (sag.) as wide (tr.) (range = 74–78%; n = 2). Anterior border increasingly prominent medially. Cranidium 93% as wide (tr.) across palpebral lobes as long (sag.) (range = 91–95%; n = 4) in palpebral view. Axial furrow anteriorly curves adaxially, intersecting lunette; there positioned 67% of width apart (tr.) as at posterior margin (range = 58–73%; n = 3). Lunette oval, less well impressed on external than on internal mould, with mid-length opposite posterior part of palpebral lobe in dorsal view. Axial furrows smoothly diverging anterior of lunette; terminating at omphalus where glabella is 138% width (tr.) at lunettes (range = 131–145%; n = 3). Omphalus very distinct on internal mould, less distinct externally; circular, with clear median granule. Anterolateral internal pit distinct on internal moulds, situated slightly closer to anterior border than to omphalus and a little farther adaxially than omphalus. Occipital node pronounced on both internal and external moulds, approximately one-tenth total sagittal length of cranidium away from posterior margin.

S1 largest glabellar muscle impression, oval (elongated exsag.), extending from midpoint of lunette to anterior of palpebral lobe. S2 smallest, transversely oval, slightly anterior of palpebral lobe. S3 a little larger than S2, transversely oval, slightly posterior of omphalus. S2 closer to S1 than S3; S1 and S2 equidistant from sagittal line; S3 slightly further abaxially.

Sculpture of pronounced, subequally spaced, continuous terrace ridges over roughly anteriormost quarter of cranidium. Cuticle pitted elsewhere.

Palpebral lobe comprising roughly one-quarter of total cranidial sagittal length. Posterior branch of facial suture directed backwards from ε, almost in line with ε with a slight outward deflection mid-course, when close to posterior margin, rapidly projects abaxially to intersect posterior margin along an exsagittal line just abaxial to abaxial limit of visual surface.

Librigena most convex medially; lateral margins with rim, this becoming more pronounced anteriorly. Short, stout genal spine most convex medially. Genal spine comprising roughly 15% of total cephalic sagittal length. Lateral border an upturned rim; very thin (sag.) border furrow. Pronounced terrace ridges branching from posterior of genal spine and short terrace ridges trending obliquely from librigena lateral margin. Remainder of librigena and area between terrace ridges covered in fairly evenly spaced pits.

Rostral plate of low convexity; most convex where upturned to meet anterior margin. Connective sutures with slight curvature, diverging strongly forwards. Only one specimen shows ventral view, with hypostomal suture represented by slightly forwardly convex median embayment. Fairly regularly spaced, continuous, transversely trending terrace ridges run parallel to anterior margin; a few short, fainter terrace ridges present between these.

Pygidia ranging in sagittal length from 10.4 to 12.5 mm. Pygidium smoothly rounded posteriorly, 54% as high as long (sag.) and almost uniformly convex in sagittal profile with maximum height approximately halfway along sagittal length (in extended view). Length (sag.) 90% of width (tr.) (range = 87–93%; n = 5) when plotted, the pygidial transverse width = 1.00 pygidial sagittal length +1.25 (see Remarks for discussion, and Fig. 16). Posterior sagittal carina variably developed; can be present for 65% total sagittal pygidial length. Abaxial end of articulating facet located roughly one-quarter total pygidial length (sag.) from anterior margin. A single pygidium (Fig. 17J–K) exhibits a pair of dark muscle impressions, located equidistant from sagittal line, anterior of articulating facet, close to anterior margin. Sculpture of terrace ridges; strongest near margins, trending transversely, discontinuous, fairly regularly spaced. Clear pitting between terrace ridges. Doublure extending 30% (range = 26–32%; n = 3) total pygidial sagittal length, of constant width, bearing well defined terrace ridges that are more regular and continuous posteriorly.

Remarks. Laneia enalios resembles both *Litotix* Hall, 1864 and *Ligiscus* Lane and Owens, 1982 in having genal spines. However, the cephalon is far less convex than in either of those genera and in this regard more resembles *Failleana*. The course of the posterior branch of the facial suture is like that of *Litotix*. The size, shape and positions of the lateral glabellar muscle impressions most resemble those of *Cybantyx*, as do the pronounced anterior border, curvature of the anterior margin, and relative proportions of both the cephalon and cranidium. These synapomorphies potentially define a clade comprising *Cybantyx* and *Laneia*.

The pygidium of *Laneia enalios* is of typical illaenimorph form. The most useful distinguishing characters are the convexity and length to width ratio, and in these respects, the pygidium of *Laneia* is most similar to that of *Paracybantyx*. The resulting straight-line equations for pygidial width against pygidial length initially suggest that the pygidia of *L. enalios* can be differentiated from those of *C. nebulosus* using this method; see Remarks for *C. nebulosus*.

Occurrence and distribution. Confined to Valdemar Glückstadt Land (VGL1).

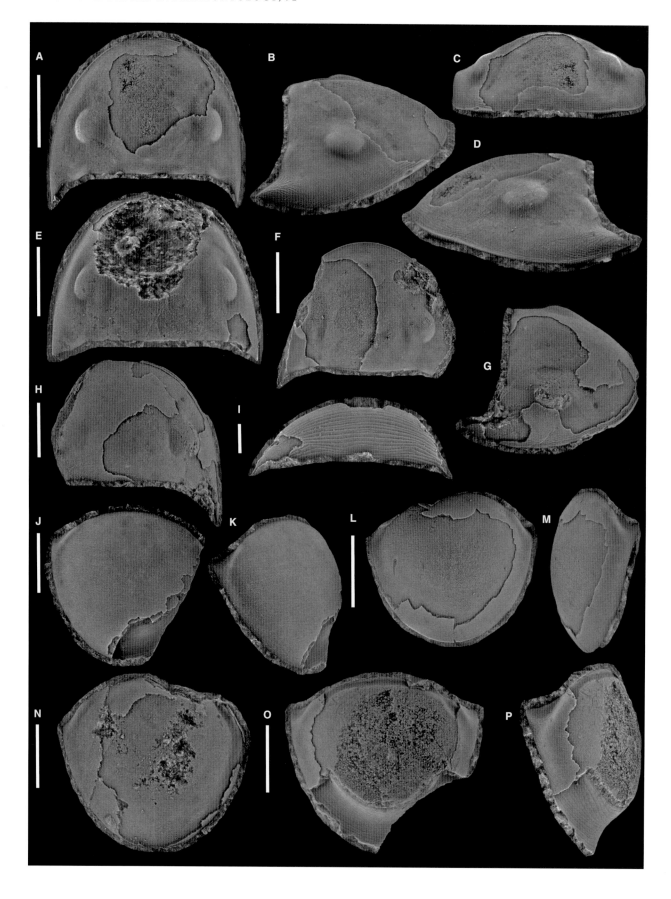

Genus LIGISCUS Lane and Owens, 1982

Type species. By original designation; *Ligiscus arcanus* Lane and Owens, 1982, from the upper Rhuddanian – Aeronian Cape Schuchert Formation, Washington Land, western North Greenland.

Other species. Ligiscus diana sp. nov.; *L. smithi* Adrain, Chatterton and Blodgett, 1995.

Diagnosis. To the diagnosis of Lane and Owens (1982, p. 46) should be added the following diagnostic characters of the pygidium, which is incompletely known in *L. arcanus*: convex pygidium, medially effaced; well developed articulated facet; first interpleural furrow most deeply impressed.

Remarks. Ligiscus is a remarkable morphological intermediate between what could be termed a typical scutelluid – with distinct dorsal furrows and a pygidium with radial ribs – and an illaenimorph. This is particularly so in the pygidium: it is highly convex, yet retains interpleural furrows and has a wide doublure with terrace ridges (in *Ligiscus*, these are distinctively scalloped-shaped). As in other taxa, effacement develops gradually through ontogeny; juvenile specimens strongly resemble non-effaced scutelluids, with more illaenimorph characters appearing with growth (compare Fig. 18A with Fig. 18L). The ontogenetic changes as exemplified by *L. diana* sp. nov., occur in other similar taxa, for example *Excetra* Holloway and Lane, 1998.

Ligiscus diana sp. nov.
Figure 18

LSID. urn:lsid:zoobank.org:act:B7446BD7-A888-4BE7-BAAE-846075D38368

Derivation of name. Latin, *Diana*, HEH's mother. Also, goddess of the chase and the moon, alluding to the cephalon which resembles a crescent moon. Noun in apposition.

Holotype. MGUH30664 (Fig. 18A–B) cephalon; from GGU 275044 (VGL2).

Figured paratypes. Locality VGL1: MGUH30666 cephalon; MGUH30667, 30669 cranidia; MGUH30674, 30676 pygidia. Locality VGL2: MGUH30665 cephalon; MGUH30668, 30670 cranidia; MGUH30671 librigena; MGUH30672 hypostome; MGUH30673, 30677–30678 pygidia.

Diagnosis. Cranidium with smoothly rounded anterior cranidial margin, and no median carina. Pygidium with shield-shaped axis, roughly as long (sag.) as wide (tr.). Non-bifid median rib.

Description. Cephala ranging in sagittal length from 1.8 to 7.3 mm. Cephalon semicircular, 53% as high as long (sag.), with maximum height opposite midpoint of palpebral lobe in palpebral view. Cephalon roughly two-thirds (64%) as long (sag.) as wide (tr.) (range = 60–69%; n = 3). Cranidium 91% as long (sag.) as wide (tr.) across midpoint of palpebral lobes (range = 87–96%; n = 25). Axial furrows subparallel adaxial to palpebral lobes; diverging posteriorly to intersect posterior margin. At lunette, axial furrows are 86% distance apart (tr.), as they are at posterior margin (range = 80–95%; n = 25). Anterior of palpebral lobes, axial furrows diverge strongly forwards, fading before anterior margin; stronger on internal mould than on external surface. Lunette visible only on internal mould; extending from midpoint of palpebral lobe to just anterior of palpebral lobe. Occipital node may be weakly developed on exterior surface. Midpoint of palpebral lobe situated approximately level with two-thirds total cephalic sagittal length from anterior margin. Anterior of axial furrows, cranidium has well defined terrace ridges following curvature of margin, mostly continuous, fairly evenly spaced (Fig. 18I).

Librigena with narrow (tr.) lateral border and border furrow; terminating in short genal spine. Sparse terrace ridges originating from posterior of visual surface.

Hypostome attributed to species with median body transversely oval, anterior wing large relative to whole hypostome, maculae small, pronounced. Covered with strong, continuous terrace ridges arranged in a 'U shape' (Fig. 18N).

Pygidia ranging in sagittal length from 3.7 to 8.6 mm. Pygidium semicircular, 55% as high as long (sag.), with maximum height approximately halfway along sagittal length (in extended view). Sagittal profile uniformly arcuate, except posteriorly where the pleural region is slightly concave; 66% as long (sag.) as wide (tr.) (range = 60–71%; n = 22); when plotted, the pygidial transverse width = 1.34 pygidial sagittal length +1.02 (see Fig. 16). Axis with faint trace of segmentation in addition to clear remnant of the first ring, comprising roughly one-third of pygidial sagittal length and roughly one-quarter of total transverse pygidial width anteriorly. Variably effaced, increasingly so with increasing size. Articulating half ring short (sag.), convex, medially bowed anteriorly. Seven pairs of pleural ribs, non-bifid median rib roughly twice as wide as paired ribs distally. First pleural rib most convex, demarcated behind by deep interpleural furrow, with articulat-

---

FIG. 17. *Laneia enalios* gen. et sp. nov. A–D, MGUH30656 holotype cephalon; A, dorsal, B, oblique lateral, C, anterior and D, lateral views. E, MGUH30657 cephalon, dorsal view. F, MGUH30658 cephalon, dorsal view. G–I, MGUH30659 cephalon; G, oblique lateral view; H, dorsal view; I, ventral view of rostral plate. J–K, MGUH30660 pygidium; J, dorsal and K, oblique lateral views. L–M, MGUH30661 pygidium; L, dorsal and M, lateral views. N, MGUH30662 pygidium, dorsal view. O–P, MGUH30663 pygidium; O, dorsal and P, lateral views. All material from GGU 275038 (VGL1). Scale bars represent 10 mm (A–H, J–P) and 1 mm (I).

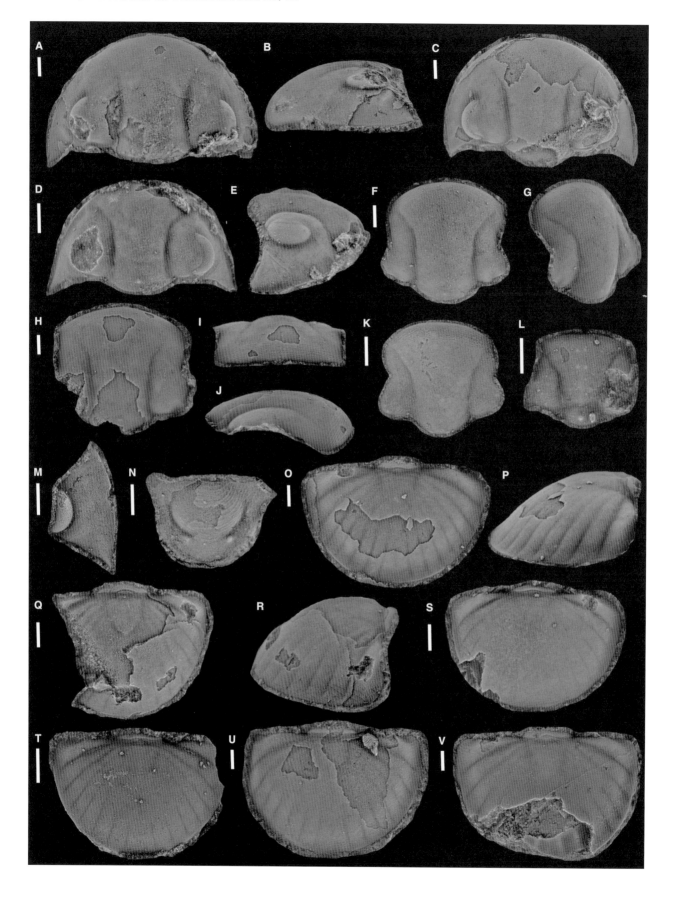

ing facet anterolaterally. Remaining paired ribs and median rib only slightly convex; all ribs widening abaxially. Interpleural furrows deepest abaxially, except the first, which retains its depth throughout its course. Pleural region variably effaced; most so medially, resulting in loss of rib convexity. Sculpture of transversely trending continuous terrace ridges; strongest on articulating facet, weakest on posteromedial part of pleural region. Doublure extending for at least one-quarter total sagittal length and bearing scalloped-shaped terrace ridges (Fig. 18V).

Remarks. Important changes with growth of this species are as follows: the cranidial axial furrow intersects with a narrow anterior border furrow in smaller specimens (Fig. 18D), whereas in larger specimens, the axial furrows fade before the anterior margin, accompanied by loss of the border (Fig. 18H); the occipital ring and furrow are only present on specimens up to 3.5 mm long (sag.) (Fig. 18L); cephala less than 3 mm long (sag.) have at least two pairs of nodes medially on the glabella (Fig. 18L); there is a proportionally longer (sag.) palpebral lobe in smaller specimens (Fig. 18D–E); and the pygidial axis exhibits gradual effacement with growth.

The cephalon of *L. diana* differs from that of the type species, *L. arcanus*, in its smoothly rounded anterior cranidial margin, lacking the anteromedial angulation shown by *L. arcanus*. The glabella is less strongly forwardly expanded than that of the type species and lacks a median carina. Cranidia of both *L. diana* and *L. arcanus* differ from those of *L. smithi* in that specimens of comparable size have axial furrows which do not intersect the anterior section of the facial suture (Adrain *et al.* 1995, p. 726, figs 2.1–2.2, 2.4–2.15, 3.13, 3.15–3.16).

The pygidium of *L. arcanus* is known from one incomplete specimen, which is effaced abaxially leaving only five discernible interpleural furrows and ribs. The pygidium of *L. diana* has a slightly greater length-to-width ratio than that of *L. arcanus*. The pygidial axes of these species are comparable in both shape and length relative to the whole pygidium. The pygidium of *L. smithi* has a similar length-to-width ratio to *L. diana*, but differs in having a bifid median rib, and the pygidial axis is wider than long and more triangular in shape, compared with

an axis as long as wide and more nearly pentagonal in *L. diana*.

Occurrence and distribution. Most abundant in Valdemar Glückstadt Land (VGL1–2). Also present in Central Peary Land (CPL1–2) and Kronprins Christian Land (KCL1, 4–5).

Genus LIOLALAX Holloway and Lane, 1999
pro *Lalax* Holloway and Lane, 1988 non Hamilton, 1990

Type species. By original designation: *Lalax olibros* Holloway and Lane, 1998, from the middle–upper Wenlock to Ludlow Mirrabooka Formation, Orange district, New South Wales.

Other species. Lalax bandaletovi (Maksimova, 1975); *L. bouchardi* (Barrande, 1846); *L. chicagoensis* (Weller, 1907); *L. clairensis* (Thomas, 1929); *L. hornyi* (Šnajdr, 1957); *L. inflatus* (Kiær, 1908); *L. kattoi* (Kobayashi and Hamada, 1984); *L. lens* (Holloway and Lane, 1998); *L. naresi* sp. nov.; *L. xestos* (Lane and Thomas *in* Thomas, 1978); *L.? sakoi* (Kobayashi and Hamada, 1984); *L.? transversalis* (Weller, 1907).

Diagnosis. As Holloway and Lane (1998, p. 877), with removal of details of morphology of the rostral plate which is not necessarily subtriangular.

Remarks. The name *Lalax* Holloway and Lane, 1998 is a junior homonym of a Cretaceous hemipteran, and was subsequently replaced by *Liolalax* Holloway and Lane, 1999. *Liolalax naresi* sp. nov. exhibits a rostral plate morphology different from previously described species of *Liolalax*, and Holloway and Lane's (1998) diagnosis is modified to reflect this.

Liolalax naresi sp. nov.
Figure 19A–I

LSID. urn:lsid:zoobank.org:act:C8C50470-5F63-41CC-9D20-5C4693F08D6C

Derivation of name. For George S. Nares, who led the British Arctic Expedition in 1875–1876 and proved that the Arctic Basin

---

FIG. 18. *Ligiscus diana* sp. nov. A–B, MGUH30664 holotype cephalon, GGU 275044 (VGL2); A, dorsal and B, lateral views. C, MGUH30665 cephalon, GGU 275044 (VGL2) dorsal view. D–E, MGUH30666 cephalon, GGU 275038 (VGL1); D, dorsal and E, oblique lateral views. F–G, MGUH30667 cranidium, GGU 275038 (VGL1); F, dorsal and G, oblique lateral views. H–J, MGUH30668 cranidium, GGU 275044 (VGL2); H, dorsal, I, anterior and J, lateral views. K, MGUH30669 cranidium, GGU 275038 (VGL1) dorsal view. L, MGUH30670 cranidium, GGU 275044 (VGL2) dorsal view. M, MGUH30671 librigena, GGU 275044 (VGL2) dorsal view. N, MGUH30672 hypostome, GGU 275044 (VGL2) ventral view. O–P, MGUH30673 pygidium, GGU 275044 (VGL2); O, dorsal and P, lateral views. Q–R, MGUH30674 pygidium, GGU 275038 (VGL1); Q, dorsal and R, oblique lateral views. S, MGUH30675 pygidium, GGU 275038 (VGL1) dorsal view. T, MGUH30676 pygidium, GGU 275038 (VGL1) dorsal view. U, MGUH30677 pygidium, GGU 275044 (VGL2) dorsal view. V, MGUH30678 pygidium, GGU 275044 (VGL2) dorsal view. All scale bars represent 1 mm.

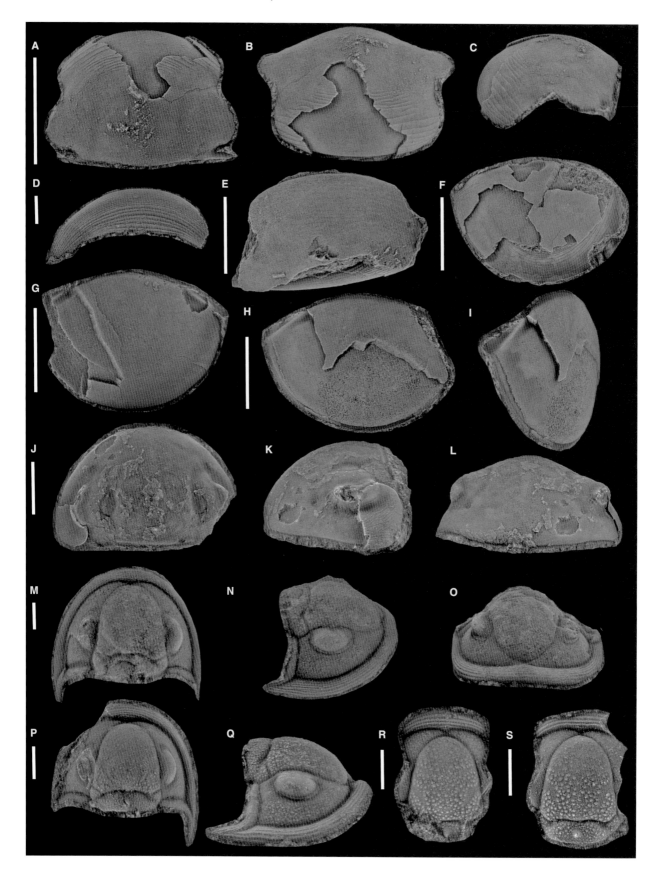

was inaccessible because of ice cover. This material was collected from South Nares land.

Holotype. MGUH30679 (Fig. 19A–C) cranidium; from GGU 298930 (SNL1).

Figured paratypes. Locality SNL1: MGUH30681 cranidium; MGUH30680 incomplete cephalic doublure; MGUH30682–30684 pygidia.

Diagnosis. Weakly impressed lunettes; cranidium about 69% as long (sag.) as wide (tr.); rostral plate lacking posterior lobe; pygidium about 65% as long (sag.) as wide (tr.); doublure present for about one-third of pygidial sagittal length; dorsal terrace ridges confined to anterior of cranidium and anterolateral margins of pygidium.

Description. The only complete cranidium is very small, being 5.6 mm long (sag.), about 19% as high as long (sag.), uniformly convex, with maximum height opposite anterior of palpebral lobe in palpebral view. Glabella slightly wider at posterior margin that at lunettes, where it comprises about 45% of width δ–δ. Lunettes very shallow and elongate (exsag.). Anterior to this, axial furrows smoothly diverging forwards until fading anterior of palpebral lobe. Axial furrows shallow throughout course. Omphalus only clear on an internal mould (Fig. 19E), where it is circular in form, and positioned much closer to facial sutures than to anterior margin, and just posterior to anteriormost glabellar muscle impressions. Glabellar muscle impressions not visible on exterior surface, and only transversely elongate S3 visible on internal mould (Fig. 19E), partly due to breakage of specimen. Palpebral lobe comprising about 36% total cranidial sagittal length, midpoint positioned about one-third total cranidial sagittal length from posterior margin in palpebral view; posterior end very close to posterior margin. Cranidial sculpture clear on exterior surface (Fig. 19A–C), comprising strongly raised, fairly continuous, non-anastomosing terrace ridges present for anteriormost 30%, more posterior ones not extending for full width of cranidium. Terrace ridges weakly visible on internal moulds. Rostral plate with numerous arcuate terrace ridges; connective sutures not clear.

Pygidia ranging in sagittal length from 6.1 to 8.5 mm; about 19% as high as long (sag.); 65% as long (sag.) as wide (tr.) (range = 62–68%; n = 3). A weak sagittal carina visible on one internal mould. Articulating facet narrow (exsag.), abaxial end located opposite roughly 20% pygidial length (sag.) from anterior. Terrace ridges only present anteriorly and anterolaterally, radiating from anterolateral extremity; weaker than those on cranidium. Doublure comprising about 33% pygidial sagittal length

and of uniform width; bearing about 12 concentric terrace ridges.

Both cranidial and pygidial cuticles are remarkably thick for size of specimens (see Fig. 19I).

Remarks. The morphology of the rostral plate of *L. naresi* is unique within the genus in lacking a posterior lobe. *L. naresi* is otherwise most similar to the type species, *L. olibros*, particularly with respect to cranidial proportions. *L. naresi* differs as follows: the lunettes are much weaker; cranidial terrace ridges confined to anterior of cranidium, whereas they are also present posteriorly on *L. olibros*; pygidium wider (tr.), of lower convexity overall, and with greatest convexity anteriorly, not posteriorly; doublure twice as long.

Lateral glabellar muscle scars have been described from cranidia of *L. lens* Holloway and Lane, 1998, from the Silurian Mirrabooka Formation, New South Wales. The S3 of *L. lens* is subcircular, as opposed to the transversely elongated form seen in *L. naresi*; both are similarly positioned in front of the omphalus, however.

Occurrence and distribution. Confined to South Nares Land (SNL1).

Liolalax? sp.
Figure 19J–L

Remarks. A cephalon measures 9.6 mm (sag.) and is almost twice as wide (tr.) as long (sag.), and 79% as high as long (sag.). A cranidium measures 10.6 mm (sag.) and is about three-quarters as long (sag.) as wide (tr.) across palpebral lobes; both are preserved as internal moulds. The occipital impression and S1 are both large and are unclear; S2 and S3 are much smaller and are transversely elongate. The specimens are tentatively assigned to *Liolalax* on the basis of the combination of the following characters: ventrally curved anterior margin, lacking rim; the presence of omphalus and anterolateral internal pit; long (exsag.) palpebral lobe; rounded genal angle. The palpebral lobe, however, is positioned slightly further from the posterior margin (its midpoint is situated about 30% total cranidial sagittal length from posterior border) and is slightly shorter (exsag.) than in other members of the genus, comprising roughly one-third total cranidial sagittal length. Additionally, the axial furrows are better

FIG. 19. A–I, *Liolalax naresi* sp. nov.; A–C, MGUH30679 holotype cranidium; A, dorsal, B, anterior and C, lateral views; D, MGUH30680 incomplete cephalic doublure, ventral view; E, MGUH30681 cranidium, dorsal view; F, MGUH30682 pygidium, dorsal view; G, MGUH30683 pygidium, dorsal view; H–I, MGUH30684 pygidium; H, dorsal and I, oblique lateral views. All material from GGU 298930 (SNL1). J–L, *Liolalax*? sp. MGUH30685 cephalon, GGU 275015 (KCL1); J, dorsal, K, lateral and L, anterior views. M–S, *Proetus*? *confluens* sp. nov.; M–O, MGUH30686, holotype cephalon; M, dorsal, N, oblique lateral and O, anterior views; P–Q, MGUH30687 cephalon; P, dorsal and Q, lateral views; R, MGUH30688 cranidium, dorsal view; S, MGUH30689 cranidium, dorsal view. All material from GGU 198225 (CPL3). Scale bars represent 5 mm (A–C, E–L) and 1 mm (D, M–S).

impressed than in species of *Liolalax*; this could reflect the preservation as an internal mould.

Occurrence and distribution. The cephalon is from Kronprins Christian Land (KCL1) and the cranidium from Valdemar Glückstadt Land (VG1).

Order PROETIDA Fortey and Owens, 1975
Family PROETIDAE Salter, 1864

Genus PROETUS Steininger, 1831

Type species. By original designation; *Calymmene concinna* Dalman, 1827, from the Wenlock Mulde Marl, Djupvik, Gotland, Sweden.

Diagnosis. See Thomas (1978, p. 36).

Remarks. Adrain (1997, p. 27) considered *Devonoproetus* Lütke, 1990 to be a subjective junior synonym of *Proetus* s.s. His argument was based on the following synapomorphies: non-vaulted, only moderately inflated glabella; long anterior border; large eye with subdued socle and narrow field; long genal spine; hypostome with a pair of posterior spines. This concept of *Proetus* has been followed by later workers, including Zhou *et al.* (2000), and Edgecombe and Wright (2004).

Proetus? confluens sp. nov.
Figures 19M–S, 20A–C

LSID. urn:lsid:zoobank.org:act:D4A1C121-0EE7-49BA-8650-3D1934E0D594

Derivation of name. Latin, *confluens*, place where two streams meet, alluding to the medially confluent preglabellar and anterior border furrows.

Holotype. MGUH30686 (Fig. 19M–O) cephalon; from GGU 198225 (CPL3).

Figured paratypes. Locality CPL3: MGUH30687 cephalon; MGUH30688–30689 cranidia; MGUH30690 articulated pygidium and thorax; MGUH30691 articulated cephalon and thorax.

Diagnosis. Glabella longer (sag.) than wide; short (sag.) preglabellar field only present abaxial to medially confluent preglabellar and anterior border furrows; anterior border expanded backwards medially; granular sculpture coarsest on glabella. Pygidium with anteriorly wide (tr.) and strongly tapering pygidial axis with at least six axial rings; at least five pleural ribs with inflated anterior and posterior pleural bands, all extending to very narrow pygidial border.

Description. Cephala and cranidia ranging in sagittal length from 3.0 to 4.0 mm (n = 9). Cephala averaging 76% as long (sag.) as wide (tr.) (range = 66–84%; n = 3). Glabella conical, excluding occipital ring averaging 80% as wide (tr.) as long (sag.) (range = 75–87%; n = 8). Axial furrows equally moderately impressed throughout their course. Occipital ring slightly wider (tr.) than L1 immediately in front and slightly longer (sag.) than anterior border, slightly convex anteriorly, and very gently sloping anteriorly; lateral lobes ovate and inflated. Occipital node situated just behind mid-length (sag.) of ring. Occipital furrow deep. Lateral glabellar muscle impressions shallow, delineated by smooth areas lacking tubercles, all connected to axial furrow. S1 with outer end more or less opposite δ, from there running posteromedially at roughly 55 degrees to an exsagittal line for a little less than half its course, after which it narrows posteriorly and is directed at a much lower angle, not reaching occipital furrow. S2 with outer end opposite γ, running posteromedially at roughly 65 degrees to an exsagittal line, widening adaxially. S3 the weakest impression, not always visible, closer to S2 than S2 is to S1, running roughly parallel to S2. Glabella densely covered in coarse granules (except for muscle impressions), fairly evenly sized and spaced, slightly finer on occipital ring than on remainder; sculpture slightly subdued towards anterior of glabella. Facial suture diverging from γ to β at roughly 12 degrees to exsagittal line; γ positioned slightly further from sagittal line than ε; posterior branch not clear. Palpebral lobe extends from just anterior of occipital lobe. Anterior border convex (sag.) with numerous parallel terrace ridges. Fixigenal field with granules of similar density to those on glabella but more subdued.

Eye comprising roughly one-third total sagittal length of cephalon. Eye socle narrow, slightly wider (tr.) anteriorly, bounded above and below by furrows. Librigenal field convex, with similar granular sculpture to fixigenal field. Lateral borders bearing numerous terrace ridges, and posterior border with terrace ridges abaxial to ω. Genal spine roughly one-quarter sagittal cranidial length, with median groove terminating before tip, where terrace ridges become confluent.

FIG. 20. A–C, *Proetus? confluens* sp. nov.; A–B, MGUH30690 articulated pygidium and thorax; A, dorsal and B, oblique lateral views; C, MGUH30691 articulated cephalon and thorax, dorsal view; both specimens from GGU 198225 (CPL3). D–H, *Airophrys balia* gen. et sp. nov.; D–F, MGUH30692 holotype cranidium; D, dorsal, E, oblique lateral and F, anterior views; G–H, MGUH30693 pygidium; G, dorsal and H, oblique lateral views; both specimens from GGU 275038 (VGL1). I–M. *Airophrys* sp. 1; I, MGUH30694 cranidium, GGU 198115 (CPL1), dorsal view; J, MGUH30695 cranidium, dorsal view; K, MGUH30696 librigena, dorsal view; L, MGUH30697 librigena, dorsal view; M, MGUH30698 cephalon, dorsal view; all material from GGU 198226 (CPL1). N–R, *Cyphoproetus* sp. 1; N–O, MGUH30699 holotype cranidium, GGU 274689 (CPL5); N, dorsal and O, lateral views; P, MGUH30700 cranidium, GGU 274689 (CPL5) dorsal view; Q, MGUH30701 cranidium, GGU 198219 (CPL2) dorsal view; R, MGUH30702 cranidium, GGU 274689 (CPL5) dorsal view. S, *Cyphoproetus* sp. 2, MGUH30703 cranidium, GGU 298924 (SNL2) dorsal view. All scale bars represent 1 mm.

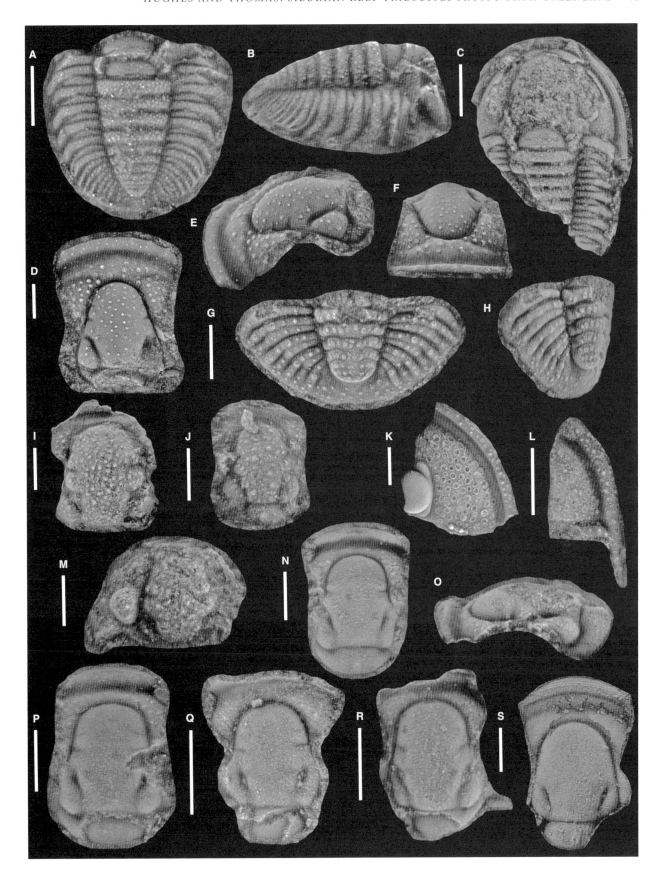

Thorax with axis gradually tapering, on average forming roughly 40% of entire thoracic width (tr.), and with deep axial furrow. Axial rings roughly equal in length exsagittally and sagittally. Pleural furrow long (exsag.), incised, terminating distally at edge of articulating facet. Axis and pleurae with similar granular sculpture.

Pygidium almost twice as wide (tr.) as long (sag.). Axis comprising roughly 40% maximum pygidial width anteriorly and roughly 70% pygidial length, tapering strongly backwards in posterior half, with strongly rounded tip. Axial furrows well impressed throughout course. Paired apodemal pits present on first four axial rings. Anterior five pleural ribs visible, posterior to this effaced. Both anterior and posterior bands equally inflated with a scalloped profile. Interpleural and pleural furrows equally well impressed, terminating at narrow pygidial border.

Remarks. This species exhibits a combination of characters which lead it to be assigned with some uncertainty to *Proetus*. The absence of a preglabellar field is typical of *Proetus*, as is the semicircular pygidium with an axis that is wide (tr.) relative to the pleural areas, and which tapers strongly. The confluent preglabellar and anterior border furrows are highly distinctive, however. A coarsely granular or tuberculate sculpture is not a character typical of *Proetus*, but it is seen in species of *Gerastos* Goldfuss, 1843, such as *G. granulatus* Lindström, 1885 (p. 81, pl. XIV, fig. 13) from the Silurian of Gotland (refigured by Owens 1973*a*, pl. 2, fig. 3) which shares similarly well defined lateral glabellar muscle impressions. Such clear lateral glabellar muscle impressions are not characteristic of *Proetus* either; the local absence of sculpture makes those impressions more conspicuous. The extension of pleural and interpleural furrows to the pygidial margin is not characteristic of the genus either.

Occurrence and distribution. Confined to Central Peary Land (CPL3).

Genus AIROPHRYS gen. nov.

LSID. urn:lsid:zoobank.org:act:D108C552-5049-4FA9-BEA0-8B6F8FA820CC

Derivation of name. Contraction of Greek, *airo*, raise, lift, and Greek, *ophrys*, brow, eyebrow, alluding to the fancied resemblance of the preocular fixigenal furrows to a pair of raised eyebrows. Gender feminine.

Type species. Airophrys balia from the Telychian Samuelsen Høj Formation, Valdemar Glückstadt Land, North Greenland.

Diagnosis. Glabella roughly as long as wide excluding occipital ring, and strongly tapering anteriorly; anterior and lateral borders raised but flat-topped; anterior and lateral border furrows deep; preocular fixigenal furrow present. Pygidium with six axial rings and a terminal piece, four pleural ribs and an anterior pleural band. No border. Pygidial axial rings, and anterior and posterior bands, all inflated. Sculpture of unevenly sized and spaced tubercles with granules between.

Remarks. Airophrys resembles *Cyphoproetus*, the genus which is probably most closely related: both have S1 deeply incised, defining a prominent L1, a similar pygidial morphology with a scalloped profile to the pleural ribs and granular sculpture. Distinguishing characters of *Airophrys* are as follows: the very strongly tapering and anteriorly narrower (tr.) frontal glabellar lobe, giving a very different glabellar outline; anterior border with a distinctly flattened profile; defined anterior border furrow; the presence of a preocular fixigenal furrow; inflated pygidial axial rings and pleural ribs. Other than the type species, *Airophrys* is only known from material described under open nomenclature: *A.* sp. 1 discussed below, and another species described as *Cyphoproetus* sp. by Lane (1972, p. 349, pl. 61, fig. 4a–b), from the Telychian Samuelsen Høj Formation of Central Peary Land and Kronpins Christians Land respectively, eastern North Greenland. In both cases, the diagnostic preocular fixigenal furrow can be seen.

Airophrys balia sp. nov.
Figure 20D–H

LSID. urn:lsid:zoobank.org:act:E2A2F5FE-4644-46A4-9D58-66F2AD4C6399

Derivation of name. Greek, *balia*, spotted, dappled, piebald, alluding to the tuberculate sculpture.

Holotype. MGUH30692 (Fig. 20D–F) cranidium; from GGU 275038 (VGL1).

Figured paratype. Locality VGL1: MGUH30693 pygidium.

Diagnosis. Glabella strongly tapering anteriorly; glabellar lobes not inflated. Unevenly spaced tubercles with very fine and dense granules between them covering entire cranidium, coarsest on preglabellar field, finest on anterior border.

Description. Glabella conical, tapering strongly forwards to a very gently rounded anterior. Glabella excluding occipital ring as long (sag.) as wide (tr.) posteriorly. S1 deep, trending at roughly 40 degrees to an exsagittal line, anterior end shallowing abruptly to meet axial furrow as a very shallow impression, posteriorly meeting occipital furrow as a deep impression. L1 subtriangular, much longer (exsag.) than wide (tr.), exsagittally comprising almost one-third the pre-occipital glabellar length. S2 and S3 equally well impressed, closer to one another than S1 is to S2; both connected to axial furrow and trending subparallel to one another at roughly 75 degrees to an exsagittal line. S3 slightly shorter than S2. Occipital ring as wide (tr.) as posterior of pre-occipital glabella; mostly incomplete. Length (sag.) of preglabellar field comprising roughly 15% total sagittal length of cranidium. Preglabellar field sloping anteriorly to meet anterior border furrow, which is well defined and shorter (sag., exsag.) medially than abaxially. Between this and the anterior border furrow is an anteriorly sloping area which adaxially becomes the preglabellar field. Anterior border with a flat-topped profile. Facial suture from γ to β diverging from exsagittal line at roughly 35 degrees.

Pygidium a little more than twice as wide as long (sag. length = 51% tr. width). Axis comprising 33% maximum pygidial width anteriorly and about 80% pygidial length, bluntly rounded posteriorly. Axial furrow well impressed throughout. Six axial rings and a terminal piece. Posteriorly, inter-ring furrows not connected medially. Paired subcircular apodemal pits connected to anteriormost five anterior ring furrows. Anteriormost pleural band slightly wider (exsag.) than those behind. Pleural furrows more deeply impressed than interpleural furrows which increase in depth towards pygidial margins, terminating shortly before margins. No distinct pygidial border. Complete covering of coarse, unevenly sized and spaced tubercles with fine granules between them.

Occurrence and distribution. Valdemar Glückstadt Land (VGL1).

Airophrys sp. 1
Figure 20I–M

Remarks. Fragmentary material comprising a cephalon, two cranidia and three librigenae from Central Peary Land are assigned to *Airophrys*. These fragmentary specimens were found in association and share a similar sculpture. Determinable characters are as follows: glabella conical, roughly as long (sag.) as wide (tr. posteriorly) excluding occipital ring; deep S1, meeting occipital furrow posteriorly; well impressed S2, trending roughly 75 degrees from an exsagittal line; posteriorly placed occipital node; short (sag.) preglabellar field with preocular fixigenal furrow; flattened lateral border, a little wider (tr.) than lateral border furrow; distinct eye socle, anteriorly and posteriorly expanded (Fig. 19K); elongate genal spine; coarse tubercular sculpture with fine granules between, with tubercles smaller and sparser on flattened lateral border and border furrow.

The material has a similar glabellar morphology to *Airophrys balia*; the pre-occipital glabella is as long (sag.) as wide (tr.) in both *A.* sp. 1 and *A. balia*. *A.* sp. 1 differs from *A. balia* in having inflated glabellar lobes, and a more coarsely granular sculpture on the glabella. A sepa-

rate species is probably represented, but the material is insufficient for it to be named and diagnosed.

Occurrence and distribution. Central Peary Land (CPL1).

Genus CYPHOPROETUS Kegel, 1927

Type species. Subsequently designated by Přibyl (1946a, p. 15); *Cyphaspis depressa* Barrande, 1846; from the Wenlock Liteň Formation, Lištice, near Beroun, Prague district, Czech Republic.

Other species. Cyphoproetus bellus (Cooper and Kindle, 1936); *C. binodosus* (Whittard, 1938); *C. comitilis* Clarkson and Howells, 1981; *C. depressus* (Barrande, 1846); *C. externus* (Reed, 1935); *C. facetus* Tripp, 1954; *C. glaudii* Lamont, 1948; *C. insterianus* Schrank, 1972; *C. latifrontalis* Schrank, 1972; *C. nasiger* (Marr and Nicholson, 1888); *C. pugionis* Howells, 1982; *C. punctillosus* (Lindström, 1885); *C. putzkeri* Šnajdr, 1976; *C. rotundatus* (Begg, 1939); *C. strabismus* Owens, 1973a.

Diagnosis. See Thomas (1978, p. 40).

Cyphoproetus sp. 1
Figure 20N–R

Description and remarks. Cranidia ranging in sagittal length from 2.1 to 2.4 mm have a pear-shaped glabella, which is bluntly rounded anteriorly. L1 is over twice as wide (tr.) as long (exsag.), comprising at least 40% total sagittal length of pre-occipital glabella, and is exsagittally expanded posteriorly where glabella is at its widest (tr.). S2 and S3 are connected to the axial furrow, and are positioned closer together than S1 is to S2. S2 is deeper than S3. S2 is directed between 65 and 85 degrees posteriorly and S3 between 75 and 85 degrees anteriorly from an exsagittally directed line. The occipital ring is over twice as long medially as abaxially and slightly wider (tr.) than glabellar width across L1. The glabellar morphology and well developed L1 are typical of *Cyphoproetus*. The absence of a tropidium or tropidial ridges distinguishes *C.* sp. 1 from species of *Warburgella* – the only other genus with a broadly similar cephalic morphology – although these characters may be weakly expressed in some *Warburgella* species, as in *W. scutterdinensis* Owens, 1973a (see Owens 1973a, text-fig. 9).

The cranidia are closest to those of *C. svartoensis* Owen (1981, p. 25, pl. 6, figs 3–6) from the Ordovician (Ashgill, late Rawtheyan) of the Oslo District, Norway. Both species lack lateral occipital lobes, have posteriorly situated occipital nodes, and share the anteriorly deepening and widening (tr.) form of S1, which is not connected to the axial furrow. The cranidium of *C.* sp. 1 differs from *C. svartoensis* in having a more elongate glabella (roughly

90% as wide as long excluding occipital ring (n = 4)) and a shorter (sag.) preglabellar field (which is shorter (sag.) than the anterior border and about equal in length (sag.) to the anterior border furrow). Although both species share a combination of both Bertillon pattern and granular sculpture, the granules of *C. svartoensis* are transversely elongate and situated on both the glabella and preglabellar field. The granules of *C.* sp. 1 are coarser, and situated only on the preglabellar field and fixigenae.

Occurrence and distribution. Central Peary Land (CPL2, 5).

Cyphoproetus sp. 2
Figure 20S

Remarks. A single cranidium measuring 3.6 mm has the following distinguishing characters: tongue-shaped glabella, about 90% as wide (tr.) posteriorly as long (sag.), excluding occipital ring; S1 the only visible glabellar furrow, and not connected to either axial or occipital furrows; two concentric furrows on the preglabellar region; posteriormost furrow is a pseudo-border furrow, dividing the preglabellar field into two equally long (sag.) sections. Preglabellar field almost twice length (sag.) of anterior border; anterior border with prominent terrace ridges anteriorly; posteriorly positioned occipital node. The deep S1 is characteristic of *Cyphoproetus*. The presence of a pseudo-border furrow on the preglabellar region, the glabellar morphology, and effaced S2 and S3 distinguish this cephalon from those assigned to *C.* sp. 1, and *Cyphoproetus* sp. 2 is not closely comparable with any described *Cyphoproetus* species.

Occurrence and distribution. South Nares Land (SNL1).

Genus OWENSUS gen. nov.

LSID. urn:lsid:zoobank.org:act:AE4BC2F5-16CF-4FCB-B5A4-3E724F8F0B49

Derivation of name. For Dr R. M. Owens, studier of proetaceans, arbitrarily combined in the style of *Proetus*. Gender masculine.

Type species. Owensus arktoperates from the Telychian Samuelsen Høj Formation, Central Peary Land, North Greenland.

Diagnosis. Highly convex cephalon, preglabellar field convex longitudinally and long (sag.); glabella not significantly increasing in width (tr.) between S1 and S2, adaxially forked S1; pygidium with seven axial rings and at least five pleural ribs with abaxially inflated pleural bands; anterior bands weakly inflated, posterior pleural bands strongly inflated; pleural and interpleural furrows equally well impressed and all extending to meet narrow border; axial region covered in coarse granules.

Owensus arktoperates sp. nov.
Figure 21

LSID. urn:lsid:zoobank.org:act:85CC6210-AD6D-4D5B-9753-34DFA04FDEBE

Derivation of name. Combination of Greek, *arktos*, northern, and Greek, *perates*, wanderer. Noun in apposition.

Holotype. MGUH30704 (Fig. 21A–B) articulated specimen; from GGU 274689 (CPL5).

Figured paratypes. Locality CPL2: MGUH30709 cephalon. Locality CPL3: MGUH30708 cranidium. Locality CPL4: MGUH30707 cranidium. Locality CPL5: MGUH30706 articulated cephalon and thorax; MGUH30705 cephalon; MGUH30712 librigena; MGUH30710–30711 articulated pygidia and thoraces.

Diagnosis. As for genus.

Description. Cephala and cranidia range in sagittal length from 1.9 to 4.9 mm, cephala averaging 74% as long (sag.) as wide (tr.) (range = 64–89%; n = 38). Glabella conical, pre-occipital part averaging 84% as wide (tr.) as long (sag.) (range = 65–101%; n = 57). Anterior to L1, tapering forwards to form a smoothly rounded point. Axial furrows well impressed. Glabellar furrows all connected to axial furrow, S1 to S3 progressively weaker. S1 well impressed, abaxial end placed opposite halfway between δ and ε, from there running posteriorly at roughly 40 degrees from an exsagittal line, until approximately one-tenth the glabellar width (tr.) at this point, where it forks for approximately another one-tenth the glabellar width. Anterior branch trending in a sagittal direction; posterior branch directed strongly posteriorly, terminating well before occipital furrow. S2 and S3 close together, lying either side of a transverse line through γ. S2 runs posteriorly at roughly 70 degrees, and S3 at roughly 80 degrees from an exsagittal line. S3 only visible on internal mould (Fig. 21F–G). Occipital ring slightly wider (tr.) than portion of glabella across

FIG. 21. *Owensus arktoperates* gen. et sp. nov. A–B, MGUH30704 holotype articulated specimen, GGU 274689 (CPL5); A, dorsal and B, lateral views. C–E, MGUH30705 cephalon, GGU 274689 (CPL5); C, dorsal, D, anterior and E, lateral views. F, MGUH30706 articulated cephalon and thorax, GGU 274689 (CPL5) dorsal view. G, MGUH30707 cranidium, GGU 274654 (CPL4) dorsal view. H–I, MGUH30708 cranidium, GGU 198225 (CPL3); H, dorsal and I, lateral views. J, MGUH30709 cephalon, GGU 198188 (CPL2) lateral view. K, N, MGUH30710 articulated pygidium and thorax, GGU 274689 (CPL5); K, dorsal and N, lateral views. L, O, MGUH30711 articulated pygidium and thorax, GGU 274689 (CPL5); L, dorsal and O, oblique lateral views. M, MGUH30712 librigena, GGU 274689 (CPL5) dorsal view. All scale bars represent 1 mm.

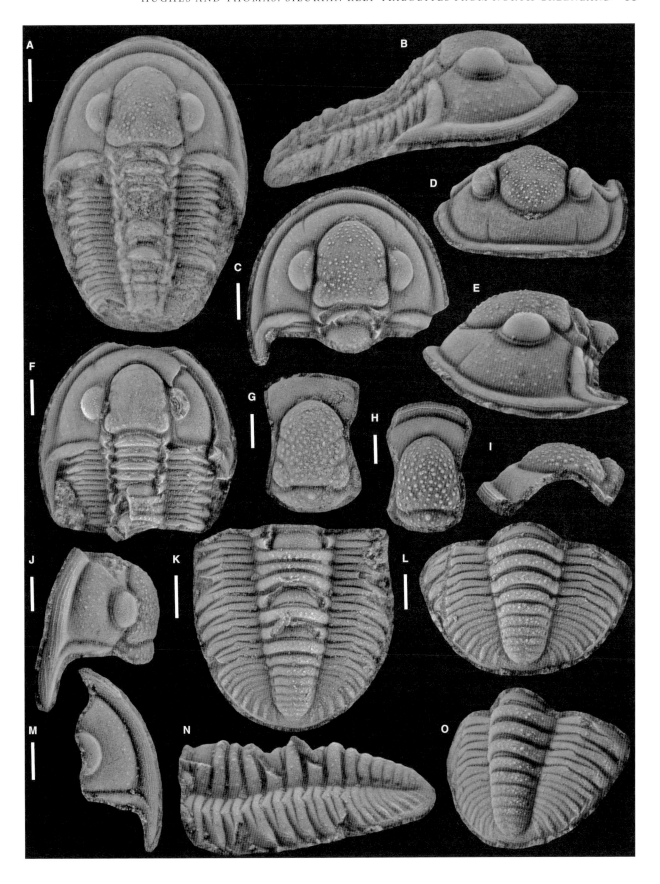

L1; convex forwards adaxially, where roughly twice as long (sag.) as abaxially. Medially, occipital ring is longer (sag.) than anterior border; anterior slope more gradual than posterior slope. Occipital furrow deep. Small ovate lateral lobe defined by shallow furrow. Pronounced occipital node situated closer to anterior than to posterior of occipital ring. Glabella densely covered in unevenly sized and spaced coarse granules. Facial suture from γ to β initially diverging from an exsagittal line at 20–30 degrees, halfway along course changing direction to run in an almost exsagittal line. γ and ε equally very close to axial furrow, and from ζ to ε running close to and almost parallel with axial furrow. Palpebral lobe situated half its exsagittal length from posterior border furrow, comprising roughly 40% total glabellar sagittal length excluding occipital ring (more in the very smallest specimens). Posterior of palpebral lobe situated opposite mid-length of L1, and anterior situated opposite L3. Length of preglabellar field on average 21% total sagittal length of cranidia (range = 15–27%; n = 57), steeply sloping anteriorly to meet deep, narrow (sag., exsag.) anterior border furrow. Remainder of genal field strongly convex also. Anterior border pronounced, convex (sag.) and with about four subparallel terrace ridges. Genal field with scattered coarse granules and finer granules and pits in between. Coarser granules are much sparser and more subdued than on glabella. Narrow eye socle retaining equal width along its length, with lower and upper margins defined by furrows, of which the lower is wider (tr.). Lateral border of similar width (tr.) to anterior border length (sag.), of similar convexity, and bearing about four terrace ridges. Posterior border with no terrace ridges, long (exsag.) abaxial of ω, rapidly narrowing adaxial of ω. Lateral and posterior border furrows deep, joining to run down genal spine as a median groove, but terminating before tip of genal spine. Genal spine roughly half sagittal length of cranidium, extending back to sixth thoracic segment and with numerous terrace ridges trending roughly exsagittally.

Thorax of 10 segments. Axis anteriorly comprising almost half total width (tr.) of thorax, tapering posteriorly to roughly one-third total width. Last ring roughly 70% width (tr.) of first ring. Axial rings slightly longer exsagittally than sagittally. Intra-annular furrow present. Axial furrow deep. Axis with sparse granules, similar to those on librigena. Nearly flat-topped pleurae with anterior and posterior pleural bands roughly equal in length (exsag.). Pleural furrows incised, longest (exsag.) at mid-width, terminating abaxially before tips of pleurae which are angular in form. Fulcrum located roughly halfway (tr.) across pleura. Pleurae with granular sculpture.

Pygidium 47% as long (sag.) as wide (tr.) (range = 43–50%; n = 6). Anterior width of axis comprising on average 31% total anterior width (tr.) of pygidium (range = 21–36%; n = 6). Axis comprising on average 85% total sagittal length of pygidium (range = 79–90%; n = 3), bluntly rounded posteriorly. Axial furrow well impressed, slightly shallower around posterior of axis. Seven axial rings and a terminal piece, inter-ring furrows becoming progressively shallower posteriorly. Paired, well impressed subcircular muscle impressions obvious on anteriormost five axial rings, with the most anterior two connected to the intra-annular and first inter-ring furrows; posterior to this, placed increasingly centrally (exsag.) on axial rings. Pygidial border bearing irregular terrace ridges.

Remarks. The high convexity cephalon with a long (sag.) preglabellar field, and pygidium with five pleural ribs with weakly inflated anterior bands, and strongly inflated posterior pleural bands, extending to meet a narrow pygidial border, serve to distinguish *Owensus* from all other proetine genera.

The long (sag.) preglabellar field with convex lateral profile of *O. arktoperates* is characteristic of some Lower Devonian genera (*Podoliproetus* Šnajdr, 1980, p. 92, *Unguliproetus* Erben, 1951, p. 95 and *Pragoproetus* Šnajdr, 1977, with emended diagnosis in Šnajdr 1980, p. 98), but *O. arktoperates* has a proportionately shorter (sag.) anterior border. The pygidium of *O. arktoperates* is distinct in that all of the pleural and interpleural furrows extend to the margin. Owens (pers. comm. 2010) suggested that *O. arktoperates* could be derived from *Proetus ainae* from the Boda Limestone Formation (Ashgill), Dalarna, Sweden (Owens 1973b, p. 122, fig. 1A–F, H–L) which was transferred to *Astroproetus* by Owens (2006, p. 135). *O. arktoperates* certainly shares most pygidial characters with *A. ainae*, although it may be that these are largely superficial: a significant difference is that in *O. arktoperates*, the posterior pleural bands are more strongly inflated than the anterior pleural bands, whereas in *A. ainae*, the reverse is true. Additionally, the pygidium of *A. ainae* is proportionally slightly longer, and has significantly shallower pleural and interpleural furrows. The cranidia are less similar, although the glabellar outlines are comparable. *O. arktoperates* has a longer (sag.) preglabellar field, better defined anterior border furrow and L1, and granular sculpture.

A single pygidium referred to *Proetus (Lacunoporaspis)* sp. from the Cape Schuchert Formation (Rhuddanian–Aeronian, B1, Idwian stage or slightly older), Washington Land, North Greenland (Lane 1979, p. 17, pl. 3, fig. 3) also shares some similarities with *O. arktoperates*: seven axial rings, mostly with subcircular muscle impressions and five pleural ribs; well impressed pleural and interpleural furrows which all extend to pygidial margins. Lane's pygidium has a wider axis (tr. relative to width of anterior of pygidium) than that of *O. arktoperates*, and the anterior bands are very slightly more inflated distally than the posterior ones; therefore, the pygidia are not considered congeneric.

Occurrence and distribution. Confined to Central Peary Land (CPL2–5).

Subfamily CRASSIPROETINAE Osmólska, 1970

Remarks. Owens (2006) revised the Subfamily Crassiproetinae, including the genera *Astroproetus*, *Boliviproetus*,

Crassiproetus, Dechenellurus, Elimaproetus, Mannopyge, Monodechenella, Raerinproetus, Thebanaspis and *Winiskia* which he regarded as a monophyletic group, likely derived from the Proetinae during the Ordovician.

Genus ASTROPROETUS Begg, 1939

Type species. By original designation; *Proetus (Astroproetus) reedi* Begg, 1939, p. 37, from the Ordovician Upper Drummuck Group (Ashgill Series, Rawtheyan Stage), Lady Burn, near Girvan, Scotland.

Other species. Astroproetus achabae (Chatterton and Ludvigsen, 2004); *A. ainae* (Warburg, 1925); *A. asteroideus* (Begg, 1939); *A. bellus* (Maksimova, 1962); *A. interjectus* (Reed, 1935); *A. owensi* (Tripp, 1980); *A. pseudolatifrons* (Reed, 1904); *A. scoticus* (Reed, 1941); *A.? enodis* (Maksimova, 1955).

Diagnosis. See Owens (2006, p. 135).

Remarks. Lütke (1990, p. 46) noted the close relationship between *Astroproetus* and *Thebanaspis*. This relationship was supported by Owens (2006, p. 135), and *Astroproetus* was included in the Crassiproetinae rather than the Tropidocoryphinae, to which it was assigned previously.

Astroproetus franklini sp. nov.
Figure 22A–I

LSID. urn:lsid:zoobank.org:act:F31512FC-4B1C-48AA-A2B2-801841CAC088

Derivation of name. For Sir John Franklin, who led an ill-fated expedition to the high Arctic in 1845. All of the material here was collected from the Franklinian Basin.

Holotype. MGUH30713 (Fig. 22A–B) cephalon; from GGU 198226 (CPL1).

Figured paratypes. Locality CPL1: MGUH30717 librigina; MGUH30718 cranidium. Locality CPL2: MGUH30714 cranidium; MGUH30715 cephalon. Locality VGL1: MGUH30716 cephalon.

Diagnosis. Glabella longer than wide, preglabellar field approaching one-fifth sagittal length of cranidia.

Description. Cephalon semicircular, averaging 60% as long (sag.) as wide (tr. at widest point) (range = 52–65%; n = 3). Glabella with maximum width at occipital ring; tapering most strongly anterior of S2 to a subtriangulate form; excluding occipital ring, as long as or slightly longer (sag.) than wide (tr.). Axial furrows well impressed throughout. Lateral glabellar muscle impressions distinct on internal mould, with S1 running from axial furrow at roughly 60 degrees from an exsagittal line for about half its

course, after which at a much lower angle, terminating in front of occipital furrow. S2 running subparallel to and extending the same distance as the first half of S1. Occipital ring slightly wider (tr.) than posterior of L1. Occipital furrow deep, becoming longer (exsag.) abaxially. Convex forwards adaxially where slightly longer (sag.) than anterior border. Occipital lobes weakly defined anteriorly; ovoid. Pronounced occipital node situated in front of mid-length of ring. Facial suture from γ to β initially diverging from an exsagittal line at roughly 40 degrees. γ and ε equally very close to axial furrow. Posterior branch diverging backwards. Palpebral lobe comprising roughly one-quarter to one-third total sagittal length of cranidium; posterior end positioned just anterior of occipital furrow. Cranidial width (tr.) averaging 60% cranidial sagittal length (range = 57–63%; n = 3). Preglabellar field longer (sag.) than anterior border. Anterior border furrow deep. Pronounced anterior and lateral borders bearing at least four distinctive, regularly spaced terrace ridges. Sculpture of fine, scattered granules, except on anterior border.

Eye socle narrow; upper and lower margins incised, retaining equal width (or height) along eye. Lateral border furrow wider (tr.) than anterior border furrow length (sag.) and shallowing posteriorly. Lateral border of similar width (tr.) to anterior border length (sag.), of similar convexity and bearing about four terrace ridges. Posterior border narrower than lateral border and lacking terrace ridges. Lateral and posterior borders joining posteriorly to bound a median groove on genal spine, this terminating before its tip. Genal spine with at least four terrace ridges, extending for over half sagittal length cranidium. Genal field with sculpture of fine scattered granules.

Remarks. The glabellar morphology, tapering to a subtriangular point anteriorly, is characteristic of *Astroproetus*, as exemplified by *A. interjectus* (Reed, 1935; see Owens 1973a, pl. 11, figs 13–15, pl. 12, fig. 1a–b), *A. scoticus* (Reed, 1941; see Owens 1973a, pl. 11, figs 5–8, 10) and *A. achabae* (Chatterton and Ludvigsen, 2004, pl. 74, figs 2, 4, 6, 8–9). These species all have a shorter (sag.) preglabellar field than *A. franklini*. *A. franklini* differs from the type species, *A. reedi*, in the following: glabella longer (sag.) than wide (tr.), rather than as wide (tr.) or a little wider than long (sag.); longer preglabellar field; granular sculpture and distinct terrace ridges on cephalic borders (the apparent absence of these from *A. reedi* could be preservational). The glabellar proportions and long preglabellar field are sufficient to distinguish *A. franklini* from all other species of *Astroproetus*.

Occurrence and distribution. Central Peary Land (CPL1, 2) and Valdemar Glückstadt Land (VGL1).

Astroproetus sp.
Figure 22J–L

Remarks. A fragmentary cephalon measures roughly 60% as long (sag.) as wide (tr.). The conical glabella, tapering

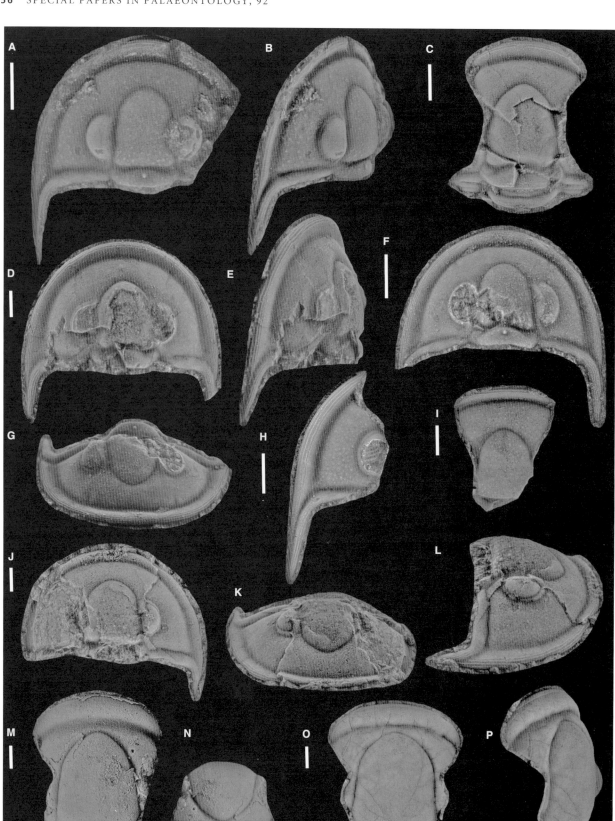

anteriorly to a subtriangular point, long (sag.) preglabellar field (comprising roughly 17% total cranidial sagittal length), anterior and lateral borders with distinct terrace ridges, position of the palpebral lobe, and genal spine morphology are similar to *Astroproetus franklini*, and the cranidium is considered congeneric. It differs from *A. franklini* in having a proportionally wider (tr.) glabella and an anteriorly expanded eye socle.

Occurrence and distribution. Kronprins Christian Land (KCL4).

Genus THEBANASPIS Lütke, 1990

Type species. By original designation; *Thebanaspis thebana* Lütke, 1990, from the Rhuddanian Edgewood Formation, Thebes, IL, USA.

Other species. Thebanaspis amblymetopa Owens, 2006; *T. determinata* (Foerste, 1887); *T. haverfordensis* (Owens, 1973*a*); *T. planedorsata* (Schmidt, 1894); *T.? arsenaulti* (Chatterton and Ludvigsen, 2004); *T.? asaphoides* (Curtis, 1958).

Diagnosis. See Owens (2006, p. 129).

Remarks. Owens (2006, p. 129) considered *Thebanaspis* to be derived from *Astroproetus*. He noted that *Thebanaspis* can be distinguished by a broader, more ovate glabella, a shorter preglabellar field, and a longer subparabolic pygidium with more axial rings and pleural ribs. The cranidial differences are obvious between the species of *Thebanaspis* and *Astroproetus* described here.

Thebanaspis sp.
Figure 22M–P

Remarks. Two cranidia with sagittal lengths of 7.5 and 8.5 mm have the subacuminate frontal glabellar lobe and short preglabellar field (roughly two-fifths length (sag.) of anterior border) characteristic of *Thebanaspis*. *Thebanaspis* sp. differs from the type species, *T. thebana*, most notably in having a relatively longer (sag.) glabella (excluding the occipital ring, it is roughly 87% as wide (tr.) as long (sag.)) with straighter, more parallel-sided axial furrows for its posterior three-quarters. Additionally, the lateral

glabellar impressions are more effaced than those of *T. thebana*.

Occurrence and distribution. Wulff Land (WL) and Western Peary Land (WPL1).

Genus WINISKIA Norford, 1981

Type species. By original designation; *Winiskia perryi* Norford, 1981, from the Telychian – ?lower Wenlock Attawapiskat Formation, Ekwan River, Ontario, Canada.

Other species. Winiskia eruga sp. nov.; *W. leptomedia* sp. nov.; *W. lissa* Owens, 2006; *W. sextaria* Owens, 2006; *W. stickta* sp. nov.

Diagnosis. See Owens (2006, p. 132), but a well defined border may be present on the pygidium, as in *W. stickta* sp. nov.

Remarks. Owens (2006, p. 132) noted the close relationship between *Winiskia* and *Thebanaspis* and stated that the genera may be distinguished by the presence of an intramarginal zone in *Winiskia*, accompanied by a more elongate glabella.

A distinctive feature of the species of *Winiskia* from the Llandovery of North Greenland and the Telychian – ?lower Wenlock Attawapiskat Formation, Ontario (as shown by *W. perryi*) is the presence of either a tapering raised area posterior of the pygidial axis or a postaxial ridge. By contrast, forms from the Llandovery of Sweden and Norway described and figured by Owens (2006) have the pygidial axis sharply defined posteriorly.

Winiskia eruga sp. nov.
Figure 23A–I

LSID. urn:lsid:zoobank.org:act:E3697A5B-566D-456E-9A6D-90DACB42AEB9

Derivation of name. Latin, *erugo*, clear of wrinkles, smooth, alluding to the loss of characters by effacement.

Holotype. MGUH30722 (Fig. 23A–B) cranidium; From GGU 274654 (CPL4).

FIG. 22. A–I, *Astroproetus franklini* sp. nov.; A–B, MGUH30713 holotype cephalon, GGU 198226 (CPL1); A, dorsal and B, oblique lateral views; C, MGUH30714 cranidium, GGU 274645 (CPL2) dorsal view; D–E, MGUH30715 cephalon, GGU 274645 (CPL2); D, dorsal and E, oblique lateral views; F–G, MGUH30716 cephalon, GGU 275038 (VGL1); F, dorsal and G, anterior views; H, MGUH30717 librigena, GGU 198226 (CPL1) dorsal view; I, MGUH30718 cranidium, GGU 198226 (CPL1) dorsal view. J–L, *Astroproetus* sp. MGUH30719 cephalon, GGU 274996 (KCL4); J, dorsal, K, anterior and L, oblique lateral views. M–P, *Thebanaspis* sp.; M–N, MGUH30720 cranidium, GGU 301326 (WPL1); M, dorsal and N, anterior views; O–P, MGUH30721 cranidium, GGU 298506 (WL); O, dorsal and P, oblique lateral views. All scale bars represent 1 mm.

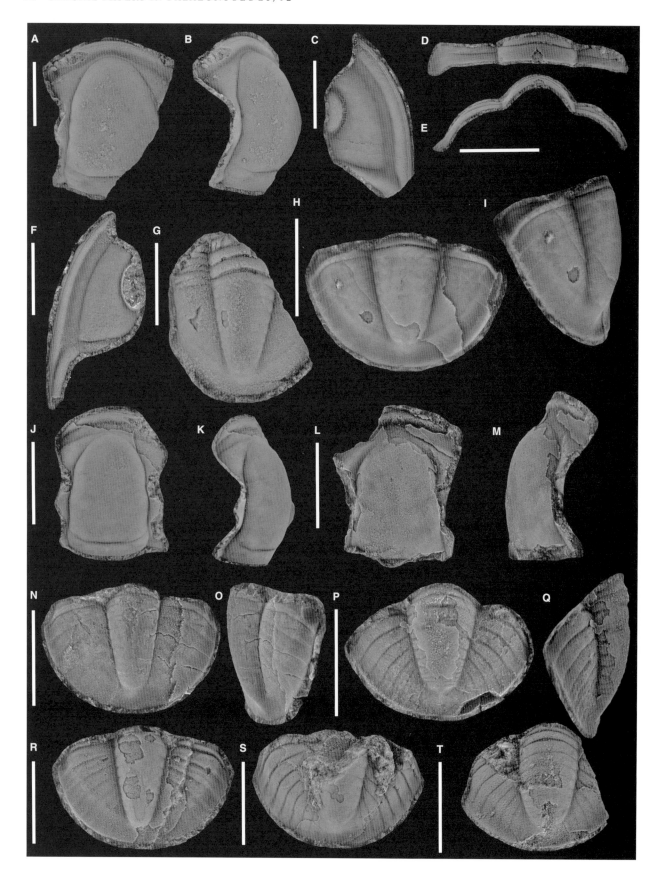

Figured paratypes. Locality CPL3: MGUH30723 librigena; MGUH30724 thoracic segment; MGUH30726 articulated pygidium and partial thorax; MGUH30727 pygidium. Locality KCL3: MGUH30725 librigena.

Diagnosis. Highly effaced glabella and pygidium, with glabella strongly tapering forwards and intramarginal zone narrow anteriorly. Sculpture of faint pits present on cephalon. Pygidium with a concave marginal zone.

Description. Glabella excluding occipital ring roughly 75% as wide (tr.) posteriorly as long (sag.); conical, bluntly rounded anteriorly. Lateral glabellar muscle impressions effaced. Posterior of palpebral lobe situated one-third entire glabellar sagittal length from posterior. Axial furrows well impressed throughout course. Occipital ring a little wider (tr.) than portion of glabella immediately in front; occipital furrow deep medially, shallowing abaxially. Occipital ring equally long (sag., exsag.) adaxially and abaxially (roughly twice sagittal length of anterior border). Lateral occipital lobes ovoid, transversely elongate, and largely effaced. Extremely short (sag.) preglabellar field present. Anterior border abaxially trending posteriorly at angle of roughly 70 degrees from sagittal line. Intramarginal zone rapidly increasing in length (exsag.) abaxially.

Narrow eye socle with well incised lower furrow. Lateral border with two terrace ridges marginally. Intramarginal zone sharply defined at both border and epiborder furrows, wide (tr.) and shallow. Lateral border faintly pitted adaxial of terrace ridges. Posterior border furrow deeper than lateral and epiborder furrows. Genal spine comprising at least one-quarter entire length (exsag.) of librigena.

Thoracic axis very convex transversely. Articulating half ring roughly half the length (sag.) of axial ring. Pleural furrow trending transversely, well defined for over half width (tr.) of pleura.

Pygidium about two-thirds as long (sag.) as wide (tr.). Axis almost straight-sided and conical; smoothly rounded and tapering posteriorly. Raised area tapering in postaxial region. Axis comprising 38% maximum pygidial width anteriorly (n = 1) and about 87% pygidial sagittal length (n = 2). Anteriormost 6 axial rings visible, posterior to this highly effaced; exact number indeterminable but possibly 9 or 10, with a terminal piece. First axial ring raised relative to others; convex sagittally as well as transversely, other rings flattened sagittally. Anteriormost pleural band clear, number of pleural ribs estimated as 7; very shallow pleural and interpleural furrows visible on anterior pleural ribs, but posterior to this, they become increasingly effaced. Concave marginal zone broad, especially posteriorly.

Remarks. The cranidium of *W. eruga* differs from the type species *W. perryi*, and all other *Winiskia* species, in the anteriorly narrow intramarginal zone. The principal difference between the cranidia of *W. eruga* and *W. perryi* is the more bluntly rounded frontal glabellar lobe of *W. perryi*, which extends to the outer edge of the intramarginal zone (the cranidium of *W. eruga*, has a short (sag.) preglabellar field). Other differences include a proportionally wider (sag.) glabella in *W. eruga*, and a more posteriorly positioned palpebral lobe than in *W. perryi*. The pygidium of *W. eruga* is relatively wider (tr.) and more effaced than that of the type species; axial rings and pleural and interpleural furrows are less clear, and it lacks a distinct postaxial ridge.

Winiskia eruga shares similarities with *W. lissa* Owens, 2006 (p. 133, fig. 7.4–7.7) from the Llandovery of Gotland, although these are considered largely superficial in nature, resulting from effacement. Both species have a very narrow band of preglabellar field medially, lateral glabellar muscle impressions effaced and weak occipital lobes. The species have different glabellar morphologies; the glabella is proportionally longer in *W. eruga*, and is less rounded anteriorly. With respect to glabellar outline, that of *W. eruga* is most comparable with that of *W.* sp. 1 (Owens 2006, p. 133, figs 7, 8a–b): both taper strongly anteriorly. The pygidium of *W. eruga*, has similar proportions to that of *W. lissa*, but differs in being more strongly effaced: in some specimens, the pleural ribs are almost invisible. In addition, *W. eruga* has a more strongly upturned margin, and the posterior of the axis is not sharply defined as in *W. lissa*: there is a raised area posterior of the pygidial axis in *W. eruga*, probably representing an effaced postaxial ridge.

Occurrence and distribution. Central Peary Land (CPL3–4) and Kronprins Christian Land (KCL3, 5).

Winiskia leptomedia sp. nov.
Figure 23J–O, R

LSID. urn:lsid:zoobank.org:act:BCF47DE2-9604-46E3-B330-B6F50CED4121

Derivation of name. Contraction of Greek, *leptos*, thin, fine, small, slender, subtle, delicate, and Latin, *medius*, middle, alluding to the narrow pygidial axis.

FIG. 23. A–I, *Winiskia eruga* sp. nov.; A–B, MGUH30722 holotype cranidium, GGU 274654 (CPL4); A, dorsal and B, oblique lateral views; C, MGUH30723 librigena, GGU 198225 (CPL3) dorsal view; D–E, MGUH30724 thoracic segment, GGU 198225 (CPL3); D, dorsal and E, lateral views; F, MGUH30725 librigena, GGU 275021 (KCL3) dorsal view; G, MGUH30726 articulated pygidium and partial thorax, GGU 198225 (CPL3) dorsal view; H–I, MGUH30727 pygidium, GGU 198225 (CPL3); H, dorsal and I, oblique lateral views. J–O, R, *Winiskia leptomedia* sp. nov.; J–K, MGUH30728 holotype cranidium, J, dorsal and K, oblique lateral views; L–M, MGUH30729 cranidium; L, dorsal and M, lateral views; N–O, MGUH30730 pygidium, N, dorsal and O, oblique lateral views; R, MGUH30731 pygidium, dorsal view. All material from GGU 275015 (KCL1). P–Q, S–T, *Winiskia* sp.; P–Q, MGUH30732 pygidium, GGU 274689 (CPL5); P, dorsal and Q, lateral views; S, MGUH30733 pygidium, GGU 274689 (CPL5) dorsal view; T, MGUH30734 pygidium, GGU 274654 (CPL4) dorsal view. All scale bars represent 5 mm.

Holotype. MGUH30728 (Fig. 23J–K) cranidium; from GGU 275015 (KCL1).

Figured paratypes. Locality KCL1: MGUH30729 cranidium; MGUH30730–30731 pygidia.

Diagnosis. Short (sag. and exsag.) occipital ring; intramarginal zone subequal in length (sag.) to anterior border; pygidial axis comprising less than one-third pygidial transverse width anteriorly.

Description. Glabella excluding occipital ring, 74% as wide (tr.) at widest point, opposite palpebral lobes, as long (sag.); tapers strongly in anterior 30% to a blunted point. S1–S3 largely effaced; faint impressions trend from axial furrows between 40 and 50 degrees from exsagittal line. Occipital ring wider (tr.) than portion of glabella immediately in front, roughly equal in transverse width to part of glabella opposite palpebral lobes, and comprising 11% total glabellar sagittal length. Occipital ring equally long (sag.) medially and at axial furrows (exsag.); occipital node placed sagittally midway across occipital ring. Anterior border convex, a little longer (sag.) than intramarginal zone. Extremely short (sag.) preglabellar field present. Posterior of palpebral lobe situated roughly one-quarter entire glabellar length (sag.) from posterior.

Pygidium averaging 65% as long (sag.) as wide (tr.) (range = 60–70%; n = 4). Axis straight-sided; smoothly rounded and tapering posteriorly. Raised area tapering in postaxial region. Axis comprising 28% maximum pygidial width anteriorly (range = 27–30%; n = 4) and on average 88% total pygidial sagittal length (range = 86–90%; n = 4). Axial rings highly effaced medially and posteriorly; exact number indeterminable, possibly nine with a terminal piece. Axial rings sagittally flattened. Anteriormost pleural band clear, and anteriormost pleural ribs distinguishable; interpleural furrows largely effaced, and pleural furrows shallow, intersecting smoothly concave marginal zone. Posterior to this, pleural and interpleural furrows increasingly effaced. Number of ribs estimated as seven.

Remarks. The cranidium of *W. leptomedia* has a similar glabellar outline to both the type species *W. perryi* (which has no preglabellar field) and *W. lissa.* It is distinguished from those species by a more posteriorly placed palpebral lobe, and differs from all other *Winiskia* cranidia by its short (sag. and exsag.) occipital ring. Cranidia of *W. leptomedia* and *W. eruga* are quite different: *W. eruga* has a more strongly tapering glabella, and shorter (sag.) intra-

marginal zone. The pygidium of *W. leptomedia* is very similar to that of *W. eruga,* but differs in its relatively narrower axis.

Occurrence and distribution. Kronprins Christian Land (KCL1).

Winiskia stickta sp. nov.
Figure 24A–K

LSID. urn:lsid:zoobank.org:act:C592B477-2C37-4F62-B5FF-F1A984AA09E3

Derivation of name. Greek, *sticktos,* punctured, spotted, dappled, alluding to the pitted sculpture.

Holotype. MGUH30735 (Fig. 24A–B) cranidium; from GGU 198214 (CPL2).

Figured paratypes. Locality CPL2: MGUH30736 cranidium; MGUH30737 librigena; MGUH30738 hypostome; MGUH30739–30742 pygidia.

Diagnosis. Broad and deep intramarginal zone, weakly defined glabellar muscle impressions and pronounced pygidial border; pitted sculpture, with pits largest and deepest within intramarginal zone and pygidial border.

Description. Glabella excluding occipital ring 86% as wide (tr.) as long (sag.) (n = 2); conical, bluntly rounded anteriorly. Glabellar furrows S1 and S2 weak but visible, both in contact with axial furrow which is inflated adaxially at these contacts. Both S1 and S2 steeply inclined, S1 forked. Occipital ring roughly equal in width (tr.) to portion of glabella immediately in front, medially longer (sag.) than at axial furrows (exsag.) and comprising 17% total glabellar sagittal length (n = 2). Occipital furrow well impressed, occipital lobes less distinct but still conspicuous. Occipital node faint, situated sagittally midway across occipital ring. Narrow preglabellar field. Anterior border gently convex sagittally. Intramarginal zone longer (sag.) than anterior border, and with distinct pits; finer pits covering all other parts of cranidium.

Librigena with lower margin of eye socle deeply incised. Lateral border with two terrace ridges marginally. Intramarginal zone sharpest at border furrow, deep and wide (tr.) Posterior border longer (exsag.) than width (tr.) of lateral border; posterior border furrow deeper than lateral epiborder furrow. Posterior

---

FIG. 24. A–K, *Winiskia stickta* sp. nov.; A–B, MGUH30735 holotype cranidium, GGU 198214 (CPL2); A, dorsal and B, lateral views; C, MGUH30736 cranidium, GGU 198200 (CPL2) dorsal view; D, MGUH30737 librigena, GGU 198200 (CPL2) dorsal view; E–F, MGUH30738 hypostome, GGU 198193 (CPL2); E, dorsal and F, oblique lateral views; G–H, MGUH30739 pygidium, GGU 198200 (CPL2); G, dorsal and H, oblique lateral views; I, MGUH30740 pygidium, GGU 198200 (CPL2) dorsal view; J, MGUH30741 pygidium, GGU 198213 (CPL2) dorsal view; K, MGUH30742 pygidium, GGU198200 (CPL2) dorsal view. L–Q, *Dalarnepeltis brevifrons* sp. nov.; L–M, MGUH30743 holotype cephalon with two articulated thoracic segments, GGU 274654 (CPL4); L, dorsal and M, lateral views; N, MGUH30744 pygidium, GGU 198226 (CPL1) dorsal view; O–P, MGUH30745 pygidium, GGU 198226 (CPL1); O, dorsal and P, lateral views; Q, MGUH30746 pygidium with posteriormost thoracic segment, GGU 274654 (CPL4) dorsal view. All scale bars represent 1 mm.

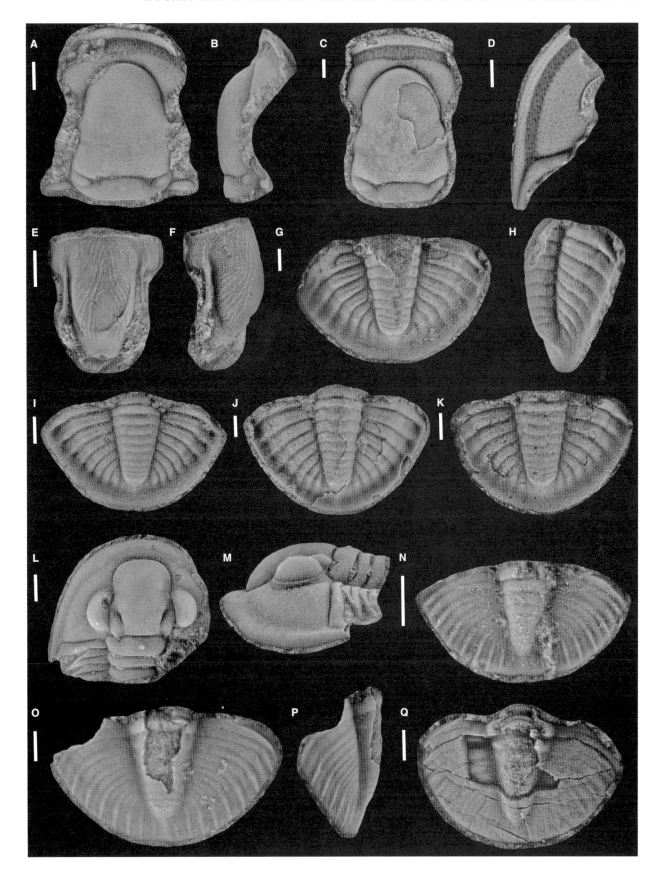

and lateral border furrows join to run down genal spine (length unknown). Entire librigena finely pitted, pits larger and deeper in intramarginal zone.

Hypostome roughly 78% as long (sag.) as wide (tr.) across anterior wings, with width across shoulders 69% that across anterior wings. Anterior wings subrectangular; roughly twice as long (exsag.) as wide (tr.). Narrow (tr.) lateral border. Middle body convex (sag. and tr.), very slender; roughly twice as long (sag.) as wide (tr.). Anterior lobe not anteriorly expanded (tr.), with lateral border furrows subparallel. Middle furrow significantly weaker than deep lateral border furrow; trending posteriorly roughly 25 degrees from exsagittal line, enclosing very narrow (tr.) maculae. Middle furrow joining lateral border furrow roughly one-third total middle body length (sag.) from posterior of middle body. Anterior lobe of middle body with terrace ridges trending from lateral border and middle furrows, converging anteriorly. Medially, where middle body lacks terrace ridges, fine pits similar to those on dorsal exoskeleton occur.

Pygidium averaging 65% as long (sag.) as wide (tr.) (range = 59–70%; n = 6). Axis straight-sided and conical; smoothly rounded posteriorly and with a postaxial ridge. Axis comprising 29% maximum pygidial width anteriorly (range = 28–29%; n = 5) and on average 79% total pygidial sagittal length (range = 75–82%; n = 6). Nine axial rings and a terminal piece. Axial rings slightly convex, those of well preserved specimens each with paired shallow, subcircular muscle impressions positioned close to the axial furrow, at mid-length (exsag.) of axial rings. Axial ring furrows well impressed. Seven pleural ribs and an anterior band. Interpleural furrows mostly effaced; faint impressions visible on anteriormost pleural ribs. Pleural furrows well impressed, anteriormost pleural furrows terminating at pronounced pygidial border; posteriorly, becoming effaced distally. Sculpture of fine pits, increasing in size and depth towards border.

Remarks. Winiskia stickta is the least effaced of the *Winiskia* species described here. The cranidium of *W. stickta* differs from that of *W. eruga* and *W. sextaria* in having a slightly longer (sag.) preglabellar field, blunter glabellar anterior, less effaced glabellar muscle impressions, a wider intramarginal zone and pitted sculpture. The pygidium of *W. stickta* is most similar to that of *W. sextaria* (Owens, 2006, p. 133, fig. 7.1–7.3) in the depth of both axial and pleural furrows and the presence of a border. The pygidium of *W. sextaria* is relatively slightly longer (sag.) than that of *W. stickta* (70% as long as wide as opposed to 65% in *W. stickta*), has an extra axial ring and pleural rib, and its pygidial border is less pronounced.

The associated hypostome (Fig. 24E–F) is the only one known from a *Winiskia* species. It is comparable to that of *Hedstroemia delicata* (Hedström, 1923 – see Owens 2006, figs 1.15–17), also included in the Crassiproetinae. The position at which the middle furrow meets the lateral border furrow, convexity of the middle body, and the position and trend of terrace ridges and fine pitting, are especially similar. The *Winiskia* hypostome lacks the strong inclination of the anterior wing away from the

median body shown by the hypostome of *H. delicata* and has the middle furrow directed at a lower angle posteriorly, resulting in more slender maculae.

Occurrence and distribution. Central Peary Land (CPL2).

Winiskia sp.
Figure 23P–Q, S–T

Remarks. Five pygidia have a straight-sided and conical axis, comprising on average 83% of the total pygidial sagittal length (n = 3), with 10 axial rings and a terminal piece. The first axial ring is raised relative to the others and is slightly convex sagittally as well as transversely, whereas the other rings are sagittally flattened. The pygidial margins are broad, especially posteriorly, and are smoothly concave. These pygidia are most comparable to that of *W. eruga*. The pygidium of *W.* sp. is distinguished by having: greater axial and pleural region convexity; more pronounced axial ring and pleural furrows, the latter of which intersect the pygidial margins (as with *W. eruga*, these do become more effaced posteriorly, and also medially); the anterior of axis relatively wider, comprising 34% of the maximum pygidial width anteriorly (range = 30–37%; n = 4); the presence of a clear postaxial ridge; exoskeleton with a finely pitted sculpture, rather than smooth.

Winiskia sp. also resembles the pygidium of *W.* sp. 1 of Owens (2006). Both are highly convex, with pleural ribs widening near the margins, show a similar degree of effacement and have sigmoidal axial ring furrows. *Winiskia* sp. differs from *W.* sp. 1 in its shorter pygidium (62% as long (tr.) as wide (exsag.) (range = 60–65%; n = 3), compared with nearly 80% as long as wide in *W.* sp. 1), upturned pygidial margins, and postaxial ridge.

Occurrence and distribution. Central Peary Land (CPL4–5).

Family TROPIDOCORYPHIDAE Přibyl, 1946*b*
Subfamily TROPIDOCORYPHINAE Přibyl, 1946*b*

Genus DALARNEPELTIS Přibyl and Vaněk, 1980

Type species. By original designation; *Decoroproetus campanulatus* Owens, 1973*b*, from the Upper Ordovician Boda Limestone Formation, Sweden.

Other species. Dalarnepeltis brevifrons sp. nov.; *D. dewingi* (Chatterton and Ludvigsen, 2004); *D. narbonnei* (Chatterton and Ludvigsen, 2004).

Diagnosis. Modified after Přibyl and Vaněk (1980, p. 166): subrectangular, waisted glabella with frontal glabellar lobe displaying an almost straight anterior margin; S1 deeply impressed, isolating oval L1; S2 and S3 visible but weaker than S1; preglabellar field short (sag.) or absent; occipital ring wider (tr.) than pre-occipital glabella, with weakly defined occipital lobes; pygidium with 7–9 axial rings, becoming increasingly effaced posteriorly, and 4–8 pairs of pleural ribs.

Remarks. Dalarnepeltis was considered by Přibyl and Vaněk (1980) to be a subgenus of *Decoroproetus* Přibyl, 1946*b*. We recognize sufficient differences from *Decoroproetus* for *Dalarnepeltis* to be considered a separate genus, particularly given the distinctive glabellar morphology. *Decoroproetus narbonnei* Chatterton and Ludvigsen (2004, p. 70, pl. 78, figs 10–17, pl. 84, figs 14–15), from the Aeronian Goéland Member of the Jupiter Formation, Anticosti Island, Québec, Canada, and *Decoroproetus dewingi* Chatterton and Ludvigsen (2004, p. 71, pl. 78, figs 1–7, 9), from the Telychian Ferrum or Pavillon Member of the Jupiter Formation, Anticosti Island, Québec, Canada, share remarkably similar glabellar morphologies, position and depth of impression of the lateral glabellar muscle impressions, and Bertillon pattern, with the type species of *Dalarnepeltis, Dalarnepeltis campanulatus* and with *Dalarnepeltis brevifrons* described below. As such, all these species should be included in *Dalarnepeltis*. The genus shows a tendency towards the shortening of the preglabellar field, which is virtually lost in *Dalarnepeltis dewingi* and *Dalarnepeltis brevifrons*.

Dalarnepeltis brevifrons sp. nov.
Figure 24L–Q

LSID. urn:lsid:zoobank.org:act:BD25015C-ED02-4F38-A40E-3F92FEB1B590

Derivation of name. Combination of Latin, *brevis*, short, and Latin, *frons*, brow, forehead, alluding to the extremely short preglabellar field. Noun in apposition.

Holotype. MGUH30743 (Fig. 24L–M) cephalon with two articulated thoracic segments; from GGU 274654 (CPL4).

Figured paratypes. Locality CPL1: MGUH30744–30745 pygidia. Locality CPL4: MGUH30746 pygidium with posteriormost thoracic segment.

Diagnosis. Cephalon with extremely short preglabellar field, shallow anterior and lateral border furrows, and δ placed opposite anterior of L1. Pygidium with eight or nine axial rings and distinct postaxial ridge; eight strongly imbricate pleural ribs with posterior pleural bands on each successively more strongly inflated abaxially.

Description. Cephalon just over 60% as long (sag.) as wide (tr.). Glabella, excluding occipital ring, almost as wide (tr.) as long (sag.). Axial furrows well impressed; from anterior of L1 converging anteriorly until S2, then subparallel-sided until near front of glabella. S1 deep, shallowing anteriorly to meet axial furrow as very shallow impression; posteriorly narrowing and shallowing, very shallow where intersecting occipital furrow. L1 comprising roughly one-third total glabellar sagittal length. S2 and S3 equally weakly impressed, both connected to axial furrow. S2 directed roughly transversely and S3 directed anteriorly at roughly 45 degrees, both for a short distance. S2 placed opposite γ, S3 half the distance from S2 as S2 is from S1. Occipital ring just the widest part of glabella, longer (sag.) than anterior border. Lateral occipital lobe only expressed anteriorly. Occipital node at mid-length of occipital ring. Occipital furrow deep, trending transversely for most of width. Facial suture from γ to β diverging from exsagittal line at 15 degrees; between β and α directed strongly adaxially; γ and ε almost in contact with axial furrow. Palpebral lobe comprising just over one-quarter total cephalic sagittal length. Posterior of palpebral lobe located just anterior of outer end of occipital furrow.

Genal field gently convex. Narrow eye socle of uniform breadth, bounded by deep upper furrow and shallower lower furrow. Anterior border length (sag.) similar to lateral border width (tr.); both very gently convex. Posterior border furrow markedly deeper than anterior and lateral border furrows, joining with lateral border furrow to run down genal spine. Genal spine length unknown. Posterior border shorter (exsag.) adaxially than lateral border width (tr.); becoming longer (exsag.) abaxially.

Bertillon pattern covering entire cephalon, most pronounced on glabella where ridges are convex forward medially.

Pygidium about 60% as long (sag.) as wide (tr.) (range = 50–65%; n = 3). Axis comprising on average 30% maximum pygidial width anteriorly (range = 25–34%; n = 3) and 88% pygidial length (range = 87–89%; n = 3), tapering to a strongly rounded terminus. Axial furrow moderately impressed; weaker around posterior of axis. Inter-ring furrows becoming very shallow posteriorly. Interpleural and pleural furrows both moderately impressed and intersecting pygidial margins; weak pygidial border.

Remarks. Dalarnepeltis brevifrons has the most effaced anterior and lateral border furrows, and the most anteriorly positioned palpebral lobe, of all *Dalarnepeltis* species. The near loss of the preglabellar field in *D. brevifrons* is shared with *D. dewingi* (Chatterton and Ludvigsen, 2004). The cranidia of these species are readily differentiated, however, by the presence of coarse granules interrupting the Bertillon pattern in *D. dewingi*, which also has a more anteriorly positioned occipital node. The pygidium of *D. dewingi* is known only from one incomplete specimen (Chatterton and Ludvigsen 2004, p. 71, pl. 78, fig. 7), but can be seen to have similar outline proportions to that of *D. brevifrons*, and a similar axial width (tr., anterior). The two species are readily differentiated by the axis of *D. brevifrons* being considerably narrower, and less rounded posteriorly, and *D. brevifrons* has a very different pleural rib structure, with imbrication increasing posteriorly (the opposite is true in *D. dewingi*), stronger interpleural furrows, and pleural furrows which are not effaced posteriorly as they are in *D. dewingi*.

The pygidium of *D. brevifrons* differs from all other species of *Dalarnepeltis* in possessing a distinct postaxial ridge: other species exhibit a tapering raised area posterior of the axis (as seen also in *Winiskia eruga* sp. nov.), which could represent an effaced postaxial ridge. *D. brevifrons* also has more axial rings and pleural ribs than the other three species. The morphology of the pleural ribs, being highly imbricate, with posterior pleural bands on each segment successively more strongly inflated abaxially, is especially similar to that of *D. campanulatus*.

Occurrence and distribution. Central Peary Land (CPL1, 4).

Tropidocoryphine? gen. et sp. indet.
Figure 25A–C

Remarks. Two pygidia are roughly twice as wide (exsag.) as long (tr.) and have an imbricate profile to the pleural ribs, a diagnostic character of the subfamily (see Owens 1973*a*, p. 40).

The number of axial rings (estimated as six or seven) and pleural ribs (there are four pleural ribs and an anterior band) agree with the familial assignment. The pygidial morphology exhibits a combination of characters which are not characteristic of any described tropidocoryphine genus, however: anterior pleural bands inflated and with coarse scattered granules, posterior pleural bands not inflated and with fewer granules; axis roughly 80% entire pygidial sagittal length, posteriorly bluntly rounded; paired apodemal pits are visible on the anteriormost rings; axial furrow well impressed, although less so around posterior of axis; interpleural and pleural furrows equally impressed, shallower in posteriormost ribs; pygidial margins concave, adaxial to a narrow pygidial border.

Occurrence and distribution. Central Peary Land (CPL2).

Order AULACOPLEURIDA Adrain, 2011
Family AULACOPLEURIDAE Angelin, 1854

Aulacopleurid cf. *Songkania* Chang, 1974
Figure 25D

Remarks. A single fragmentary cephalon from Valdemar Glückstadt Land has the following distinguishing characters: subquadrate glabella; isolated L1; short, almost transverse S2; long preglabellar field; long (sag.) anterior border furrow. The glabella morphology and length of the preglabellar field are reminiscent of species assigned to *Songkania* (see *S. smithi* Adrain and Chatterton, 1995, fig. 5).

Occurrence and distribution. Valdemar Glückstadt Land (VGL1).

Family SCHARYIIDAE Osmólska, 1957

Diagnosis. See Owens and Fortey (2009, p. 1215).

Remarks. Owens and Hammann (1990) included only *Scharyia* in the Subfamily Scharyiinae and assigned the subfamily to the Family Brachymetopidae. They reassigned *Panarchaeogonus* Öpik, 1937, to the Family Aulacopleuridae, based on the following characters: isolated L1; over 10 thoracic segments; short, transverse pygidium. Adrain and Chatterton (1993) argued that an isolated L1 is a condition general to the Aulacopleurida and that the only species of *Panarchaeogonus* for which the number of thoracic segments is known, does in fact contain nine (*P. trigodus* Warburg, 1925). Additionally, they noted that although poorly known, the pygidium of *Panarchaeogonus* is quite long relative to its width, like that of *Scharyia* but unlike that of aulacopleurids. Adrain and Chatterton (1993) returned to the shared characters of *Scharyia* and *Panarchaeogonus* recognized by Owens (1974) and considered the following indicative of relationship: triangulate glabella; large palpebral lobe; position of the eye; occipital ring morphology. Adrain and Chatterton (1993) therefore considered the Scharyiidae as a separate family of the Aulacopleurida, including the genera *Scharyia*, *Panarchaeogonus* and also *Niuchangella* Chang, 1974, which they considered to have affinities to *Panarchaeogonus*. Adrain and Kloc (1997, pp. 707–709) considered *Scharyia* likely to form a clade with *Panarchaeogonus* and *Proscharyia*, and followed the concept of the Family Scharyiidae as in Adrain and Chatterton (1993). Adrain and Westrop (2007, pp. 338–343), in their discussion of the dimeropygid trilobites, noted that *Lasarchopyge* Chatterton, Edgecombe, Waisfeld and Vaccari, 1998, was not closely related to any genus included in the Dimeropygidae. Instead, they regarded the genus as 'an obvious member of the Scharyiidae', being very similar to *Panarchaeogonus*. Furthermore, Owens and Fortey (2009, p. 1215) described new species of *Lasarchopyge* from the Upper Ordovician of Arctic Russia, which share the following characters with *Scharyia*, supporting inclusion of both genera within the same family: glabella without isolated L1; cedariiform postocular facial suture; length and segmentation of the pygidium. Owens and Fortey (2009) agreed that *Panarchaeogonus* shares more characters with *Scharyia* than believed by Owens (*in* Owens and Hammann 1990) and noted that the two genera are linked particularly by their pygidial proportions which also differentiate them from other members of the family.

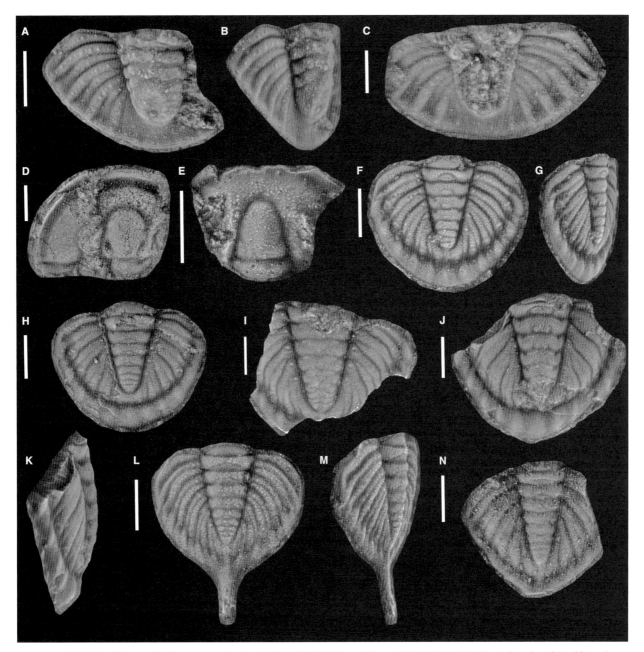

FIG. 25. A–C, Tropidocoryphine? gen. et sp. indet.; A–B, MGUH30747 pygidium, GGU 198186 (CPL2); A, dorsal and B, oblique lateral views; C, MGUH30748 pygidium, GGU 198201 (CPL2) dorsal view. D, Aulacopleurid cf. *Songkania* Chang, 1974, MGUH30749 cephalon, GGU 275038 (VGL1) dorsal view. E, *Scharyia* sp. 1, MGUH30750 cranidium, GGU 275038 (VGL1) dorsal view. F–K, *Scharyia* sp. 2; F–G, MGUH30751 pygidium; F, dorsal and G, oblique lateral views; H, MGUH30752 pygidium, dorsal view; I, MGUH30753 pygidium, dorsal view; J–K, MGUH30754 pygidium; J, dorsal and K, lateral views. L–N, *Scharyia* sp. 3; L–M, MGUH30755 pygidium; L, dorsal and M, oblique lateral views; N, MGUH30756 pygidium, dorsal view; F–N from GGU 198226 (CPL1). All scale bars represent 1 mm.

The affinities of *Proscharyia* Peng, 1990 (pl. 19, figs 7–15) from the Tremadoc Madaoyu Formation, north-western Hunan, remain unresolved. The genus was included in the Scharyiinae by Peng (1990) and Adrain and Fortey (1997), and as forming a clade at family level with *Scharyia* and *Panarchaeogonus* by Adrain and Kloc (1997). The

genus was listed as a questionable bathyurid by Jell and Adrain (2003, pp. 431, 467).

The concept of the Scharyiidae followed here is that of Jell and Adrain (2003, p. 479) and Owens and Fortey (2009), who included the following genera: *Scharyia* Přibyl, 1946*b*; *Lasarchopyge* Chatterton, Edgecombe, Waisfeld and

Vaccari, 1998; *Panarchaeogonus* Öpik, 1937; *Niuchangella* Chang, 1974.

Genus SCHARYIA Přibyl, 1946*b*

Type species. By original designation; *Proetus micropygus* Hawle and Corda, 1847, from the Ludlow Kopanina Formation, Prague district, Czech Republic.

Other species. Scharyia abdita Hörbinger, 2004; *S. angusta* Přibyl, 1966; *S. archiaciana* Přibyl, Vaněk and Hörbinger, 1985; *S. brevispinosa* Přibyl, 1967; *S. corona* Šnajdr, 1980; *S. couviniana* Osmólska, 1957; *S. dadula* Šnajdr, 1980; *S. hamlagdadica* Alberti, 1981; *S. hecuba* Šnajdr, 1980; *S. heothina* Owens, 1974; *S. hubeiensis* Xiang and Zhou, 1987; *S. kabylica* Alberti, 1981; *S. kathena* Wright and Chatterton, 1988; *S. kitabica* Feist *in* Owens *et al.*, 2010; *S. maura* Alberti, 1970; *S. meridiana* Alberti, 1970; *S. micropyga wenlockiana* Přibyl, 1967; *S. nympha* Chlupáč, 1971; *S. perscita* Přibyl, Vaněk and Hörbinger, 1985; *S. redunzoi* Perry and Chatterton, 1979; *S. ritchei* Chatterton and Campbell, 1980; *S. scharyi* Šnajdr, 1980; *S. siceripotrix* Owens, 1974; *S. sola* Šnajdr, 1976; *S. tafilaltensis* Alberti, 1970; *S. vesca* Přibyl, 1966; *S. yolkini* Přibyl, 1970; *S. yolkiniana* Alberti, 2004; *S. zeravschanica* Ivanova and Owens, 2008.

Diagnosis. See Owens (1974, p. 688).

Scharyia sp. 1
Figure 25E

Remarks. A cranidium from Valdemar Glückstadt Land is assigned to *Scharyia* on the basis of its triangulate glabella and preglabellar field which is weakly convex in longitudinal section. The glabellar outline is closest to *S. siceriptorix* Owens, 1974 (p. 689–693, pl. 98, figs 1–9) from the Lower Elton Beds (Ludlow) of the Malvern Hills and Wenlock Edge, UK: excluding the occipital ring, it is roughly as wide (tr. posteriorly) as long (sag.), and the occipital ring is a little wider (tr.) than the remainder of the glabella. Additional synapomorphies with *S. siceriptorix* include the proportions of the preglabellar field (it is half the sagittal length of the glabellar, excluding the occipital ring), the lateral occipital lobes are effaced, and the anterior section of the facial suture is strongly divergent. *S.* sp. 1 differs in having deeper border furrows, and also has an incised S1. The degree of effacement of the lateral glabellar furrows is a variable character within *Scharyia*. The moderately incised S1 in *S.* sp. 1 is most comparable with that of *S. micropyga* (Hawle and Corda, 1847) *wenlockiana* Přibyl, 1967 from the Wenlock of the Prague district, Czechoslovakia, and *S. heothina* Owens, 1974, from the Upper Ordovician Boda Limestone Formation, Sweden.

Occurrence and distribution. Valdemar Glückstadt Land (VGL1).

Scharyia sp. 2
Figure 25F–K

Remarks. Nine pygidia, with sagittal lengths ranging from 2.0 to 3.6 mm, have the small granule on the adaxial end of each posterior pleural band diagnostic of the genus. The conical, straight-sided axis is also characteristic of *Scharyia*. They are closest to a single pygidium of *S.* sp. of Lane (1972, p. 352, pl. 61, fig. 11), from the Telychian Samuelsen Høj Formation of Kronprins Christian Land, eastern North Greenland. Both species have an anterior band and five pleural ribs which are curved strongly backwards, and the overall proportions are similar; pygidia average 76% as long (sag.) as wide (tr.) (range = 72–82%; n = 7), and the axis comprises on average 70% of the entire sagittal length of the pygidium (range = 51–86%; n = 5), with the axial width comprising about one-third (34%, range = 32–35%, n = 7) of the entire pygidial transverse width. Other similarities include a fine granular sculpture, the axial morphology, having seven or eight axial rings and a terminal piece, with the depth of impression of ring furrows increasing towards a deeply impressed axial furrow, and the border furrow becoming increasingly interrupted by ribs anteriorly.

Scharyia sp. 2 differs from Lane's specimen most obviously in the presence of a border with five pairs of inflated lobes and one terminal inflated lobe, compared with the very low swellings present in his specimen. In *Scharyia* sp. 2, both border and lobes smoothly increase in size posteriorly, and paired lobes appear connected to the posterior pleural band on each segment. Additionally, the pleural furrows of ribs four and five do not reach the axial furrow, and *S.* sp. 2 has a deeper border furrow. *Scharyia* sp. 2 also has a larger granule present on the abaxial end of each anterior pleural band (except the first), in addition to the granule present on the adaxial end of each posterior pleural band.

A single pygidium assigned to *Scharyia* sp. from the Telychian of the Holitna Group, Taylor Mountains, southwestern Alaska (Adrain *et al.* 1995, p. 732, fig. 5.15), possesses similarly inflated marginal lobes to *S.* sp. 2. These are more elongated and spine-like than the marginal lobes of the Greenland pygidia. The Alaskan species is too poorly preserved to enable further comparison. *Scharyia* sp. B, represented by a single fragmentary pygidium, from the Ludlow Tamchi Regional Stage, Chaltash Formation, Shaly Sai, northern Nuratau Range, Uzbekistan (Ivanova *et al.* 2009, p. 730, fig. 10.2), also displays pronounced nodes on the pygidial border, but unlike the equivalent swellings of *S.* sp. 2, these are clearly isolated. This pygidium also differs from those of *S.* sp. 2 in having swollen abaxial ends of the anterior pleural bands. The two largest pygidia of *S.* sp. 2 measure 3.5 and 3.6 mm (Fig. 25I–K) and exhibit differences from the smaller pygidia which are considered onto-genetic rather than intraspecific in nature: the course of the interpleural furrows is more sigmoidal; border lobes are significantly more inflated further posteriorly; both pleural and interpleural furrows are less well impressed; pleural ribs are more distinct from the border; granular sculpture is both finer and more dense.

Occurrence and distribution. Central Peary Land (CP1).

Scharyia sp. 3
Figure 25L–N

Remarks. Two subtriangular pygidia with a granular sculpture, averaging 82% as long (sag. excluding spine) as wide (tr.), have an axis comprising 79% total pygidial sagittal length, with eight axial rings and a terminal piece. At its anterior end, the axis comprises 32% of the entire pygidial width. The strongly tapering axis is closest in form to that of *S. kitabica* Feist, 2010, from the Devonian of southern Uzbekistan. *S.* sp. 3 has a very distinctive pleural region, with very strongly backwards curving pleural and interpleural furrows, and pleural ribs exhibiting a marginal inflection, creating the appearance of a border furrow, with ribs inflated abaxial to this, creating a border. The postaxial ridge merging with the border, extends into a terminal spine, reminiscent of that of *Dalmanites* Barrande, 1852. The end of the spine is broken, but seems to comprise roughly one-quarter entire pygidial sagittal length. This is the only species of *Scharyia* with a terminal pygidial spine.

Occurrence and distribution. Central Peary Land (CPL1).

Order PHACOPIDA Salter, 1864
Suborder CHEIRURINA Harrington and Leanza, 1957
Family CHEIRURIDAE Hawle and Corda, 1847
Subfamily CHEIRURINAE Hawle and Corda, 1847

Genus CHEIRURUS Beyrich, 1845

Type species. Subsequently designated by Barton (1916, p. 129); *Cheirurus insignis* Beyrich, 1845, p. 12, pl. (unnumbered), fig. 1; from the Wenlock Liteň Formation, Svatý Jan pod Skalou, Czech Republic.

Other species. Cheirurus centralis Salter, 1853; *C. barkolensis* Zhang and Sun, 2007; *C. dilatatus* Raymond, 1916; *C. falcatus* sp. nov.; *C. gotlandicus* Lindström, 1885; *C. hitoeganensis* Kobayashi and Hamada, 1987; *C. infensus* Campbell, 1967; *C. niagarensis* (Hall, 1867); *C. obtusatus* Hawle and Corda, 1847; *C. patens* Raymond, 1916; *C. phollikodes* Holloway, 1980; *C. postremus* Lane, 1971; *C. prolixus* Holloway, 1980; *C. shetaensis* Nan, 1976; *C. strabo* Weber, 1932; *C. strux* Alberti, 1970;

C. tarquinius Billings, 1869; *C. uratubensis* Weber, 1932; *C.? gerassimovi* Yanishevsky, 1918.

Diagnosis. See Lane (1971, p. 11).

Cheirurus falcatus sp. nov.
Figures 26, 27A–G

LSID. urn:lsid:zoobank.org:act:D67C4141-AC43-4A15-9915-EBE18E34711D

Derivation of name. Latin, *falcatus*, sickle-shaped, curved, alluding to the shape of the pygidial spines.

Holotype. MGUH30768 (Fig. 27A–B) pygidium; from GGU 274689 (CPL5).

Figured paratypes. Locality CPL2: MGUH30757–30758 cranidia; MGUH30763 hypostome; MGUH30766 thoracic segment; MGUH30769–30770 pygidia. Locality KCL5: MGUH30759, 30762 cephala. Locality CPL5: MGUH30760 cranidium; MGUH30761 glabella; MGUH30764–30765 hypostomes; MGUH30767 thoracic segment; MGUH30771 pygidium.

Diagnosis. Glabellar width across L1 comprising 30% cephalic transverse width at this point; anterior part of glabella over 20% wider than posterior part across L1; S1 shallowing posteriorly; L1 comprising 30% transverse glabellar width; S2 located the same distance from S1 as S1 is from the occipital furrow, and S3 situated at a slightly greater distance than this from S2; pygidial pleural spines decreasing in length backward.

Description. Cephala and cranidia ranging in sagittal length from 4.7 to 35.8 mm, averaging 57% as long (sag.) as wide (tr.) at posterior of lateral border (range = 56–58%; n = 5). Glabella almost straight-sided, width across L1 comprising 30% total transverse width of cranidium/cephalon (range = 28–32%; n = 6); gradually widening (tr.) towards frontal lobe. Posterior of glabella across L1, averaging 79% of its anterior transverse width (range = 71–90%; n = 11). S1 deepest abaxially; firstly trending obliquely back from axial furrow, then much more strongly at which point it narrows and shallows, reaching the occipital furrow as a shallow impression (depth of impression there variable and independent of size). S2 and S3 anteriorly convex, trending subparallel to one another; longest (exsag.) at axial furrow, becoming shorter and weaker adaxially, terminating at roughly one-third transverse glabellar width. Abaxial position of lateral glabellar furrows from occipital furrow as percentage of pre-occipital glabellar length (sag.) as follows: S1 on average 19% (range = 15–24%; n = 12); S2 on average 38% (range = 34–46%; n = 12); S3 on average 61% (range = 57–70%; n = 12). Occipital ring projecting forwards medially so that abaxially it is between half and two-thirds the sagittal length. Occipital furrow deeper abaxially. Median occipital node situated roughly halfway (sag.) along occipital ring. Glabellar

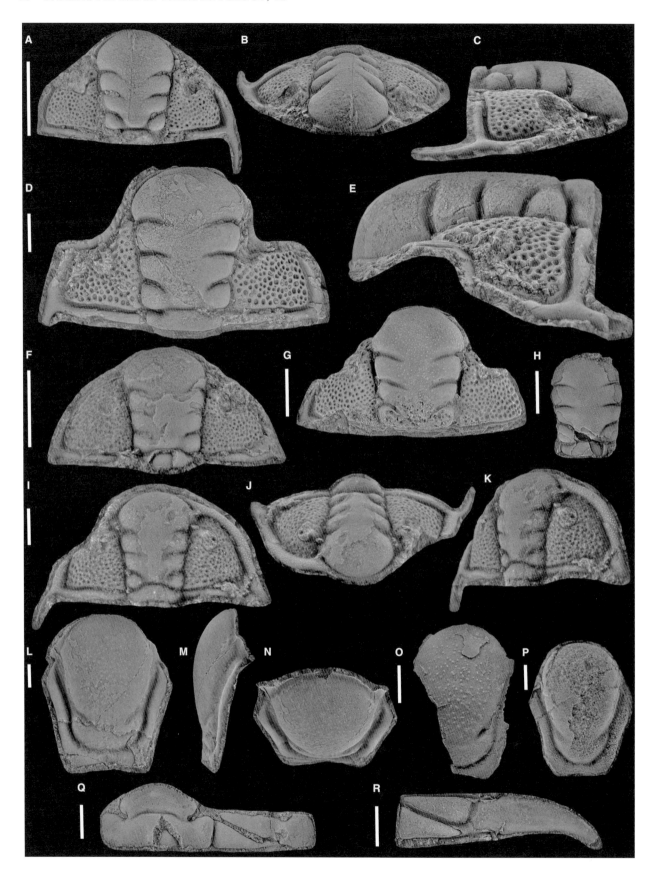

sculpture comprising small granules; very densely packed on medial and anterolateral part of frontal lobe. A few scattered slightly larger granules occur laterally on L1–L3 and on postero-lateral part of frontal lobe. Anterior border gradually shortening (sag.), and anterior border/preglabellar furrow shallowing adaxially, to merge with frontal glabellar lobe medially. Palpebral lobe placed opposite posterior half of L3. Eye ridge intersecting axial furrow just anterior of S3. Anterior branch of facial suture directed smoothly towards axial furrow, intersecting anterior border approximately at mid-length (sag.) of frontal lobe of glabella at a point very close to axial furrow. Posterior branch running in line with S2, intersecting lateral border furrow opposite posterior half of L3. From there, directed slightly posteriorly until midway across lateral border, after which directed very strongly posteriorly, meeting outer margin of border opposite L2. Genal field of low convexity between axial furrow and palpebral lobe, sloping downwards more strongly abaxial of palpebral lobe. Lateral border width (tr.) greater than posterior border length (exsag.). Lateral and posterior border furrows equally well impressed. Genal spine slender, roughly one-quarter cephalic sagittal length. Sculpture of genal field comprising unequally sized but rather equally spaced large, flat-bottomed pits which are subcircular to oval in dorsal view. Lateral and posterior borders with small, densely packed granules.

Hypostome about 80% as wide (tr.) as long (sag.) across anterior of shoulders. Middle body smoothly convex (sag. and tr.); elongate, roughly two-thirds as wide (tr.) at widest point (near anterior of middle body) as long (sag.). Anterior lobe of middle body about four times as long (sag.) as posterior lobe, with maculae placed roughly 30% from posterior of hypostome, slightly inflated with a shallow middle furrow. Anterior and posterior margins of middle body smoothly rounded. Anterior border and shallow anterior border furrow continuous medially. Lateral and posterior border furrows deep; lateral border furrow becoming progressively wider (tr.) and slightly shallower anteriorly. Lateral border of constant width; posterior border laterally expanded with angular posterolateral corners. Entire hypostome very finely granulose; middle body additionally with larger widely spaced scattered granules.

Thoracic segment with articulating half ring pronounced, almost as long (sag.) as axial ring. Articulating furrow shallow, particularly near apodemal pit. Axial furrow deep, sharp. Inner portion of pleura roughly 60% as long (exsag.) as wide (tr.); divided by deep pleural furrow into equally sized and shaped triangular anterior and posterior pleural bands. Outer portion of pleura curved backwards in most distal quarter, terminating in spinose tip.

Pygidium, including spines, roughly twice as wide (tr.) across tip of first pair of pleural spines as long (exsag. from posterior of articulating half ring to tip of last pair of pleural spines). Axis triangular, averaging 68% as wide (tr. anteriorly) as total pygidial length (sag.) (range = 65–76%; n = 6). Axis comprising three rings and a terminal piece. Axial rings increasingly convex (sag.) anteriorly. Terminal piece of axis continues posteriorly into short, rounded, posteromedian marginal projection. Axial furrow well impressed through most of course, except posteromedially. Three pairs of pleural spines; from anterior to posterior, each pair successively decreasing a little in length and increasingly backwardly projected. First pleural furrow resembling those of thoracic segments, but less posteriorly directed. Second pleural furrow shorter than first, running from anterior of pleura and terminating at its mid-length (exsag.). Third pleural furrow running from near anterior of pleura for a very short distance. Sculpture, where preserved, granulose (Fig. 27A–B).

Ontogeny. One cephalon (MGUH30762; Fig. 26I–K) is significantly smaller (3.7 mm sag. length) than other cephala and cranidia (6.3–35.8 mm sag. length) and differs from these in the following: eye placed slightly more anteriorly, the posterior margin is placed opposite the posterior half of L3, and the anterior margin opposite posterior portion of frontal glabellar lobe; genal spine more abaxially deflected; glabella proportionally narrower; S3 shorter than S2, rather than equal; confluence of lateral and posterior border furrows is angular, rather than rounded.

Remarks. Cheirurus falcatus differs from the type species, *C. insignis*, and from most *Cheirurus* species, most notably in the morphology of S1 and L1; S1 shallows posteriorly and L1 comprises under one-third the glabellar width (tr.) in *C. falcatus*. In most *Cheirurus* species, L1 is more elongated and clearly isolated by strong S1 furrows. In our material, cephala and cranidia which are otherwise identical display a variation in the strength of S1 approaching the occipital furrow (e.g. Fig. 26A cf. Fig. 26D); this variation is independent of size and therefore is not attributed to ontogeny.

The pygidium of *C. falcatus* is typical of the genus. The strength of the axial furrow, the similarity of the anteriormost pygidial pleurae to those of the thoracic segments, and the length of the posteromedial marginal projection are the most distinctive generic characters. Species of *Cheirurus* may either have their pleural spines equal in length or slightly decreasing in length backwards (Lane 1971, p. 11).

Occurrence and distribution. Most abundant in Central Peary Land (CPL1–5), also present in Kronprins Christian Land (KCL5), and Valdemar Glückstadt Land (VGL1).

FIG. 26. *Cheirurus falcatus* sp. nov. A–C, MGUH30757 cranidium, GGU 198194 (CPL2); A, dorsal, B, anterior and C, lateral views. D–E, MGUH30758 cranidium, GGU 198186 (CPL2); D, dorsal and E, lateral views. F, MGUH30759 cephalon, GGU 274774 (KCL5) dorsal view. G, MGUH30760 cranidium, GGU 274689 (CPL5) dorsal view. H, MGUH30761 glabella, GGU 274689 (CPL5) dorsal view. I–K, MGUH30762 cephalon, GGU 274774 (KCL5); I, dorsal, J, anterior and K, oblique lateral views. L–N, MGUH30763 hypostome, GGU 198198 (CPL2); L, dorsal, M, lateral and N, anterior views. O, MGUH30764 hypostome, GGU 274689 (CPL5) dorsal view. P, MGUH30765 hypostome, GGU 274689 (CPL5) dorsal view. Q, MGUH30766 thoracic segment, GGU 198198 (CPL2) dorsal view. R, MGUH30767 thoracic segment, GGU 274689 (CPL5) dorsal view. Scale bars represent 5 mm (A–H, L–N, Q–R) and 1 mm (I–K, O–P).

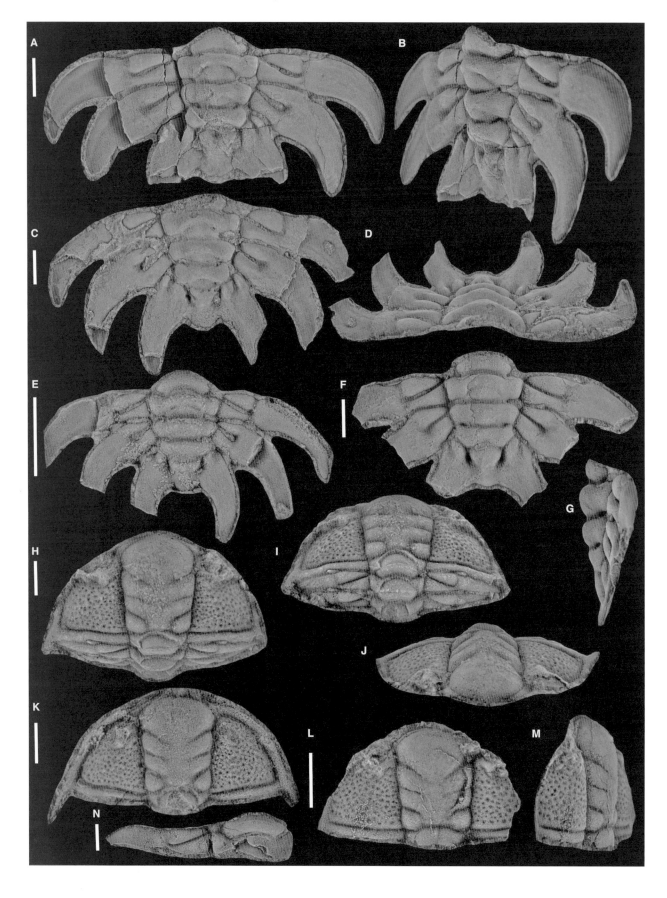

Genus PROROMMA Lane, 1971

Type species. By original designation; *Proromma bregmops* Lane, 1971, from the Lower Llandovery Skelgill Beds, Crummack Dale, near Austwick, Yorkshire.

Diagnosis. See Curtis and Lane (1997, p. 39).

Proromma? sp.
Figure 27H–N

Description. Cephala and cranidia ranging in sagittal length from 3.7 to 8.5 mm, averaging 51% as long (sag.) as wide (tr.) at posterior of lateral border (range = 47–56%; n = 4). Glabella almost straight-sided, posterior part across L1 comprising 30% total transverse width of cranidium/cephalon (range = 29–32%; n = 3); gradually widening (tr.) towards frontal lobe. Posterior of glabella across L1 averaging 88% anterior transverse width (range = 81–92%; n = 5). S1 well impressed, initially straight and directed posteromedially at 60–70 degrees from an exsagittal line for one-third glabellar width (tr.); from there abruptly shallowing to run almost exsagittally to connect with the occipital furrow. Depending on the faintness of this part of the furrow, L1 may be fully or partially isolated. S2 well impressed, trending as a straight line posteriorly from axial furrow at 70–80 degrees from an exsagittal line, for approximately one-third glabellar width (tr.). S3 well impressed, trending as a straight line posteriorly from axial furrow at roughly 60 degrees from an exsagittal line, for just over one-third glabellar transverse width. Abaxial position of lateral glabellar furrows from occipital furrow (as percentage of pre-occipital glabellar length (sag.)) is as follows: S1 on average 21% (range = 17–25%; n = 6); S2 on average 38% (range = 34–45%; n = 6); S3 on average 64% (range = 60–70%; n = 6). Hence, S2 is located slightly closer to S1 than S1 is to the occipital furrow, and S3 is the most isolated furrow. Occipital ring expanded forwards medially. Deep occipital furrow. Anterior border and preglabellar furrow absent medially. Posterior of palpebral lobe placed opposite mid-length of L3, anterior placed opposite S3. Anterior branch of facial suture directed smoothly towards axial furrow, intersecting anterior border in anterior portion of frontal lobe of glabella at a point close to axial furrow. Posterior branch running anterolaterally from posterior of palpebral lobe to intersect lateral border roughly mid-length along L3. Fixigena at highest point in line (sag.) with palpebral lobe; smoothly sloping adaxial of this. Lateral border width (tr.) subequal to posterior border length (exsag.); lateral and posterior borders equally well impressed. Genal spine directed slightly adaxial to lateral border, extending for one-quarter cephalic sagittal length. Densely packed subcircular to oval pits cover genal fields.

Thoracic axial ring width (tr.) forming roughly one-quarter entire thoracic segment width (tr.) and about twice the width (tr.) of inner portion of pleura. Articulating half ring roughly two-thirds as long (sag.) as axial ring. Axial and pleural furrows deeply impressed. Inner portion of pleura roughly 58% as long (exsag.) as wide (tr.), divided by pleural furrow into equally sized triangular anterior and posterior pleural bands. Outer portion of pleura curved backwards to terminate in spinose tip.

Remarks. The forwardly placed eye and posteriorly shallowing S1 suggest that this species could belong to *Proromma*. Without knowledge of the pygidia, the generic assignment cannot be confirmed. Previously described *Proromma* species are found in shelf-edge facies and are commonly associated with graptolite faunas, rather than in reef limestones.

Much *Proromma* material from the British Llandovery described by Lane (1971, p. 38) and Curtis and Lane (1997, p. 39) is distorted, which may in part account for the variation displayed by the lateral glabellar furrows, impeding comparison with the material described here. No other *Proromma* species, however, exhibits a perfectly straight S3, which runs at a lower angle and for a greater distance than S2. *P.?* sp. is also distinguished by having a glabella which is only 12% wider (tr.) anteriorly than posteriorly: other species display a greater anterior glabellar expansion.

Occurrence and distribution. Present in Valdemar Glückstadt Land (VGL1), Central Peary Land (CPL5) and Kronprins Christian Land (KCL1).

Genus RADIURUS Ramsköld, 1983

Type species. By original designation; *Radiurus phlogoideus* Ramsköld, 1983, from the Lower Visby Marl, Telychian (Llandovery), Norderstrand, Visby, Gotland, Sweden.

Other species. Radiurus adraini Chatterton and Ludvigsen, 2004; *R. avalanchensis* Chatterton and Perry, 1984; *R. certus* Poulsen, 1934; *R. estonicus* Männil, 1958; *R. hyperboreus* Poulsen, 1934.

Diagnosis. Modified from Ramsköld (1983, p. 188), to account for variations in length between pairs of pygidial spines, and to include the lack of a pygidial posteromedi-

FIG. 27. A–G, *Cheirurus falcatus* sp. nov.; A–B, MGUH30768 holotype pygidium, GGU 274689 (CPL5); A, dorsal and B, oblique lateral views; C–D, MGUH30769 pygidium, GGU 198213 (CPL2); C, dorsal and D, anterior views; E, MGUH30770 pygidium, GGU 198169 (CPL2) dorsal view; F–G, MGUH30771 pygidium, GGU 274689 (CPL5); F, dorsal and G, lateral views. H–N, *Proromma?* sp.; H–J, MGUH30772 cranidium with two articulated thoracic segments, GGU 275038 (VGL1); H, dorsal, I, posterior and J anterior views; K, MGUH30773 cephalon, GGU 274689 (CPL5) dorsal view; L–M, MGUH30774 cranidium, GGU 275038 (VGL1); L, dorsal and M, oblique lateral views; N, MGUH30775 thoracic segment, GGU 275015 (KCL1) dorsal view. Scale bars represent 5 mm (A–G) and 1 mm (H–N).

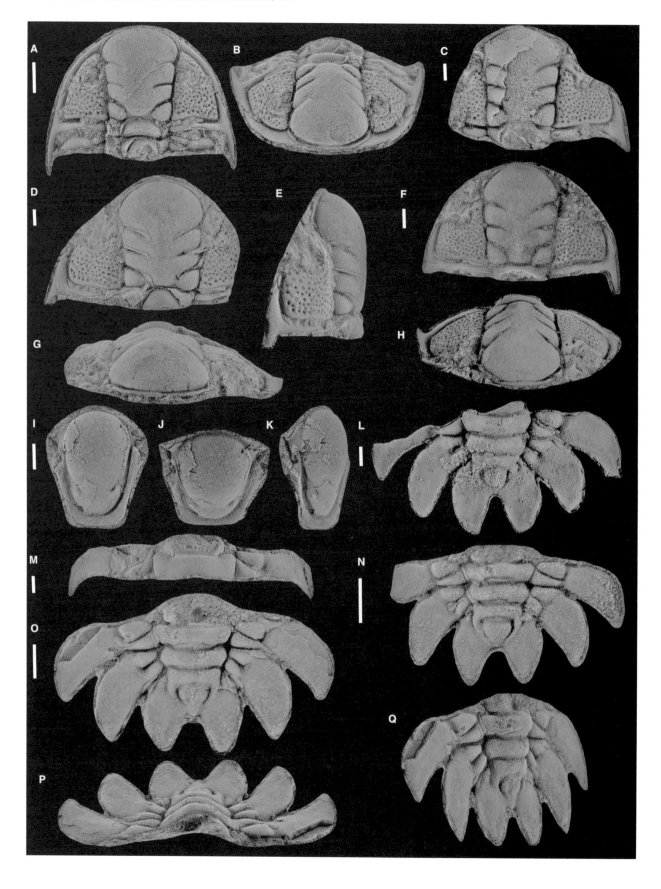

an marginal projection which all assigned species exhibit: cheirurine with entire anterior border and preglabellar furrow; S3 and S2 reaching one-third to three-eighths across glabella. Pygidium with three pairs of radially disposed, equally spaced spines; axis commonly terminating at posterior border; posteromedian marginal projection absent.

Remarks. Lane (1971, p. 77) first remarked that Silurian cheirurids with radially disposed spines may have their ancestry near *Radiurus estonicus* and *Cheirurus*. The plesiomorphic characters of *Radiurus* were commented on also by Ramsköld (1983), Norford (1994) and Chatterton and Ludvigsen (2004). Norford (1994) suggested that from *Radiurus*, a more inflated frontal glabellar lobe and development of a small pygidial posteromedian marginal projection would lead to *Cheirurus*, and forward relocation of the palpebral lobe and a pronounced pygidial posteromedian marginal projection, to *Chiozoon* Lane, 1972. All three genera are found in North Greenland. Comparison of *C. falcatus* (above), *R. pauli* (below) and *Chiozoon cowiei* Lane, 1972 from the Telychian Samuelsen Høj Formation of Kronprins Christian Land, shows that the taxa share the position of the palpebral lobe between S2 and S3, and have similarly shaped pygidial pleural spines (although these differ in length). The principal characteristics distinguishing the genera are the degree of development of the posteromedian marginal projection and the complete anterior border in *Radiurus*.

Radiurus pauli sp. nov.
Figure 28

LSID. urn:lsid:zoobank.org:act:0CB15524-A787-478A-8C59-939E2A143565

Derivation of name. For HEH's husband.

Holotype. MGUH30784 (Fig. 28O–Q) pygidium; from GGU 298506 (WL).

Figured paratypes. Locality WL: MGUH30776 cephalon; MGUH30777–30778 cranidia; MGUH30780 hypostome; MGUH30781 pygidium; MGUH30782 thoracic segment. Locality SNL2: MGUH30779 cranidium; MGUH30783 pygidium.

Diagnosis. Palpebral lobe placed between S2 and S3; pygidial pleural spines short and broad, successively decreasing in length posteriorly.

Description. Cephala and cranidia ranging up to 12.9 mm sagittal length, averaging 54% as long (sag.) as wide (tr.) at posterior of lateral border (range = 50–60%; n = 5). Glabella gradually expanding in width (tr.) anteriorly, with posterior of glabella across L1 averaging 90% of its widest point at frontal lobe (range = 77–100%; n = 10). Frontal glabellar lobe rounded; flattened with less anterior expansion in smaller specimens (Fig. 28A–B). S1 deepest abaxially; following smooth trajectory to intersect occipital furrow. L1 subtriangular, comprising roughly 33% glabellar width at this point. S2 and S3 anteriorly convex, subparallel; longest (exsag.) at axial furrow, becoming shorter and weaker adaxially until both terminate at roughly 30% glabellar width. Abaxial position of lateral glabellar furrows from occipital furrow (as percentage of pre-occipital glabellar length (sag.)) as follows: S1 on average 24% (range = 19–26%; n = 10); S2 on average 40% (range = 36–43%; n = 10); S3 on average 62% (range = 58–68%; n = 10). S2 located closer to S1 than to S3. Occipital ring expanded forwards medially; occipital furrow of constant depth. Glabellar sculpture comprising small, scattered granules. Preglabellar furrow shortest (sag.) and shallowest medially. Eye ridge intersecting axial furrow at abaxial end of S3. Anterior branches of facial suture smoothly converging anteriorly, intersecting anterior border approximately opposite mid-length (sag.) of frontal lobe of glabella, very close to axial furrow. Posterior branch running at same angle and parallel to S2, intersecting lateral border roughly opposite S2. From there, directed slightly posteriorly until midway across lateral border, after which directed very strongly posteriorly, meeting outer margin opposite S1. Fixigena convex, increasingly so abaxially. Lateral border of subequal width (tr.) to posterior border length (exsag.). Lateral border furrow broader than posterior border furrow; furrows equally well impressed. Genal spine slender, roughly one-quarter cephalic sagittal length. Sculpture of genal field comprising unequally sized and spaced pits; these subcircular to oval in dorsal view.

Hypostome about 80% as wide (tr.) as long (sag.) across anterior of shoulders. Middle body smoothly convex (sag. and tr.); ovoid, about 70% as wide (tr.) as long (sag.) at widest point near anterior of middle body. Anterior lobe of middle body about four times as long (sag.) as posterior lobe, with maculae placed roughly 35% from posterior of hypostome. Anterior and posterior margins of middle body smoothly rounded. Anterior border and anterior border furrow continuous medially. Lateral and posterior border furrows deep; lateral border furrow progressively widening anteriorly. Lateral border pinched inwards,

FIG. 28. *Radiurus pauli* sp. nov. A–B, MGUH30776 cephalon, GGU 298506 (WL); A, dorsal and B, anterior views. C, MGUH30777 cranidium, GGU 298506 (WL) dorsal view. D–E, G, MGUH30778 cranidium, GGU 298504 (WL); D, dorsal, E, lateral and G, anterior views. F, H, MGUH30779 cranidium (not blackened), GGU 298924 (SNL2); F, dorsal and H, dorsoanterior views. I–K, MGUH30780 hypostome, GGU 298506 (WL); I, ventral, J, anterior and K, oblique lateral views. L, MGUH30781 pygidium, GGU 298506 (WL) dorsal view. M, MGUH30782 thoracic segment, GGU 298504 (WL) dorsal view. N, MGUH30783 pygidium, GGU 298924 (SNL2) dorsal view. O–Q, MGUH30784 holotype pygidium, GGU 298506 (WL); O, dorsal, P, anterior and Q, oblique lateral views. All scale bars represent 1 mm.

and so narrowest (tr.) opposite maculae; posterior border of constant width with gently rounded posterolateral corners. Fine granules seen where cuticle is present.

Only available thoracic segment with axial ring comprising about one-third total segment width. Articulating half ring pronounced. Articulating furrow shallow. Axial furrow deep, sharp. Inner portion of pleura almost as long (exsag.) as wide (tr.); divided by deep pleural furrow into similarly sized and shaped triangular anterior and posterior pleural bands. Outer portion of pleura curved backwards, terminating in spinose tips.

Pygidium including spines 46% as long (exsag. from posterior of articulating half ring to tip of last pair of pleural spines) as wide (tr. across tip of first pair of pleural spines) (range = 45–50%; n = 5). Axis triangular, averaging 86% as wide (tr.) anteriorly as long (sag.) (range = 83–90%; n = 4). Axis comprising three rings and a terminal piece. Axial rings increasingly convex (sag.) anteriorly. Terminal piece of axis enclosed by ridges connected abaxially to posteriormost axial ring; reaching posterior margin. Axial furrow well impressed through its anteriormost half. Three pairs of pleural spines, each successively decreasing in length posteriorly and increasingly backwardly directed so that posteriormost pair are roughly half the length of the anteriormost pair. First pleural furrow like those of thoracic segments; second pleural furrow shorter than first, less obliquely directed than first pleural furrow.

Remarks. *Radiurus pauli* is closest to the type species, *R. phlogoideus*, especially in glabellar structure. *R. pauli* differs from *R. phlogoideus* and all other species of *Radiurus* in having a more anteriorly placed eye. The pygidial axial morphology of *R. pauli* is very similar to that of *R. phlogoideus* also, particularly regarding the morphology of the terminal piece, which is connected abaxially to the posteriormost axial ring. *Radiurus pauli* lacks the pair of tubercles on each pygidial axial ring exhibited by *R. phlogoideus*. The pygidia of *R. pauli* are readily distinguishable from *R. phlogoideus* and all other species of *Radiurus* by their short and broad pleural spines. The pleural spines of other species are more slender, and longer, particularly in *R. adraini* Chatterton and Ludvigsen, 2004 (see pl. 48, figs 8–12; pl. 59, fig. 4), from the Telychian Jupiter Formation, Anticosti Island, Québec, Canada, and *R. avalanchensis* Chatterton and Perry, 1984 (see pl. 7, figs 16–22, 24–6 and text-fig. 13), from the Llandovery Whittaker Formation (lower part of the *amorphognathoides* Biozone (Over and Chatterton 1987)), Mackenzie Mountains, north-western Canada. The pleural spines of *R. phlogoideus*, and *Cheirurus?* sp. of Norford (1981, see pl. 9, fig. 7; subsequently assigned to *Radiurus* by Ramsköld, 1983) from the Telychian – ?lower Wenlock Attawapiskat Formation, Canada, are closer to those of *R. pauli*. The pygidium of *R. pauli* is additionally differentiated from all of the above species in having pairs of pleural spines which are progressively

shorter posteriorly, so that the posteriormost pair is proportionally very short. Other *Radiurus* species have pleural spines of subequal length.

Occurrence and distribution. Most abundant in Wulff Land (WL). Also present in South Nares Land (SNL1–2) and Western Peary Land (WPL1–2).

Subfamily ACANTHOPARYPHINAE Whittington and Evitt, 1954

Remarks. Adrain (1998) provided a history of study and undertook a cladistic analysis of the Subfamily Acanthoparyphinae. The Silurian species analysed were considered to form a monophyletic group including the genera *Hyrokybe* Lane, 1972, *Parayoungia* Chatterton and Perry, 1984, and *Youngia* Lindström, 1885.

Genus HYROKYBE Lane, 1972

Remarks. *Shiqiania* Chang, 1974, originally described from the Silurian of China, is a subjective junior synonym of *Hyrokybe* (Chatterton and Perry 1984; Adrain 1998).

Type species. By original designation; *Hyrokybe pharanx* Lane, 1972 from the Telychian Samuelsen Høj Formation of Kronpins Christians Land, eastern North Greenland.

Other species. *Hyrokybe canadensis* (Billings, 1866); *H. copelandi* (Perry and Chatterton, 1979); *H. gaotanensis* (Chang, 1974); *H. globiceps* (Lindström, 1885); *H. globosa* (Yi, 1978); *H. hadnagyi* Chatterton and Perry, 1984; *H. inermis* (Lindström, 1885); *H. julli* Chatterton and Perry, 1984; *H. latisulcata* (Yi, 1989); *H. lenzi* Chatterton and Perry, 1984; *H. lightfooti* Adrain, 1998; *H. luorepingensis* (Wu, 1977); *H. meliceris* Lane and Owens, 1982; *H. mitchellae* Adrain, 1998; *H. punctata* (Chang, 1974); *H. uralica* (Chernyshev, 1893); *H. youngi* Adrain, 1998; *H.? globiceps* (Lindström, 1885).

Diagnosis. See Adrain (1998, p. 707).

Remarks. As with some other cheirurid genera, differentiating between *Hyrokybe* and *Youngia* is problematical when the hypostomal and pygidial morphology is not known. Cephalic characters at first thought to be diagnostic, for example the lack of S3 in the type species of *Hyrokybe*, were subsequently found to be unreliable. Chatterton and Perry (1984, p. 29) provided a diagnosis based on a number of species of *Hyrokybe* for which pygidia were also known. Adrain's (1998) diagnosis was composed following his cladistic analysis of the Acanthoparyphinae.

Hyrokybe pharanx? Lane, 1972
Figure 29A–C

Remarks. The proportions of the glabella, the morphology of the lateral glabellar furrows – which comprise a very distinct S1 connected to the axial furrow – and the very faint S2 and even fainter S3 (both of which are close to but isolated from the axial furrow), are all characteristic of the type species of *Hyrokybe*, *H. pharanx* Lane, 1972 (see pl. 64, figs 1–3). The cranidia are assigned to *H. pharanx* with some uncertainty as their granular sculpture is much denser than that of *H. pharanx*. The hypostome and pygidium of the species are unknown.

Occurrence and distribution. Confined to Central Peary Land (CPL2, 5).

Hyrokybe? sp.
Figure 29D–J, M

Remarks. Hyrokybe? sp. differs from *H. pharanx* most notably in the deeper S2 and S3, and the denser granular sculpture. The material is close to *H. meliceris* Lane and Owens, 1982 from the Llandovery–Wenlock of Kap Schuchert, Washington Land, western North Greenland. Particular similarities are as follows: course of the lateral glabellar furrows; convexity of the librigena; closely packed granular sculpture. *Hyrokybe*? sp. has a more distinct S2, and this could account for its observed connection with the axial furrow (S2 is distinct from the axial furrow in *H. meliceris*), and the glabella of *H.*? sp. is relatively slightly longer (sag.) than that of *H. meliceris*.

Similarities in glabellar morphology, course and depth of impression of the lateral glabellar furrows, and sculpture, are shared with *H.*? *globiceps* (Lindström, 1885; see Ramsköld 1983, pl. 26, figs 7, 10–11) from the Wenlock, Upper Visby Marl, Visby area, Gotland, Sweden. The two species are readily distinguished by the position of the palpebral lobe: the anterior margin of the palpebral lobe is situated opposite S1 in *H.*? *globiceps*, whereas the posterior margin is situated opposite S1 in *H.*? sp. Without further exoskeletal parts, certain attribution of this material to *Hyrokybe* rather than *Youngia* is not possible, and the material is not formally named. However, the course of S1 in lateral profile, with a posteriorly convex deflection about halfway up the glabella, is an apomorphic character state partly defining the node upon which the clade containing *H. julli*, *H. hadnagyi*, *H. lenzi*, *H. copelandi* and *H. youngi* is based in Adrain's (1998, p. 701, fig. 3) analysis of the Acanthoparyphinae. The presence of this apomorphy, combined with the general similarities of

H.? sp. to species of *Hyrokybe*, supports the provisional assignment to *Hyrokybe*.

Occurrence and distribution. Confined to Central Peary Land (CPL4–5).

Acanthoparyphinae gen. et sp. indet. 1
Figure 29K–L

Remarks. A single cranidium from Central Peary Land is assigned to the Subfamily Acanthoparyphinae on the basis of its subcircular glabella with maximum width across L1 and its granular sculpture. The cranidium differs from others described here, in having: a less rounded glabella; the palpebral lobe much farther from the axial furrow; S1 more effaced posteriorly, and straighter; finer, more irregularly spaced granular sculpture. The morphology of the glabella, and the eye positioned well away from the axial furrow, distinguish this cranidium from all species of *Hyrokybe* and *Youngia*: its generic assignment is uncertain.

Occurrence and distribution. Represented by a single cranidium from Central Peary Land (CPL1).

Acanthoparyphinae gen. et sp. indet. 2
Figure 29N–P

Remarks. A 4.4 mm long (sag.) granulose glabella lacking the occipital ring from Central Peary Land is characteristic of the Subfamily Acanthoparyphinae. The fragment is distinctly wider (tr.) than long (sag.) and is distinguished from all other Acanthoparyphinae here described by its coarser granular sculpture and the presence of a shallow S4 directed from the anterior margin. Generic assignment is not possible based on such limited material.

Occurrence and distribution. Represented by a single cranidium from Central Peary Land (CPL5).

Family ENCRINURIDAE Angelin, 1854
Subfamily ENCRINURINAE Angelin, 1854

Genus DISTYRAX Lane, 1988

Type species. By original designation; *Distyrax peeli* Lane, 1988, from the Telychian Odins Fjord Formation, Central Peary Land, North Greenland.

Other species. Distyrax americana (Vogdes, 1886); *D. bibullatus* sp. nov.; *D. cooperi* Edgecombe and Chatterton, 1992; *D. maccormicki* Chatterton and Ludvigsen, 2004; *D. pilistverensis* (Rosenstein, 1941); *D.*? *quinquecostata* (Männil, 1958).

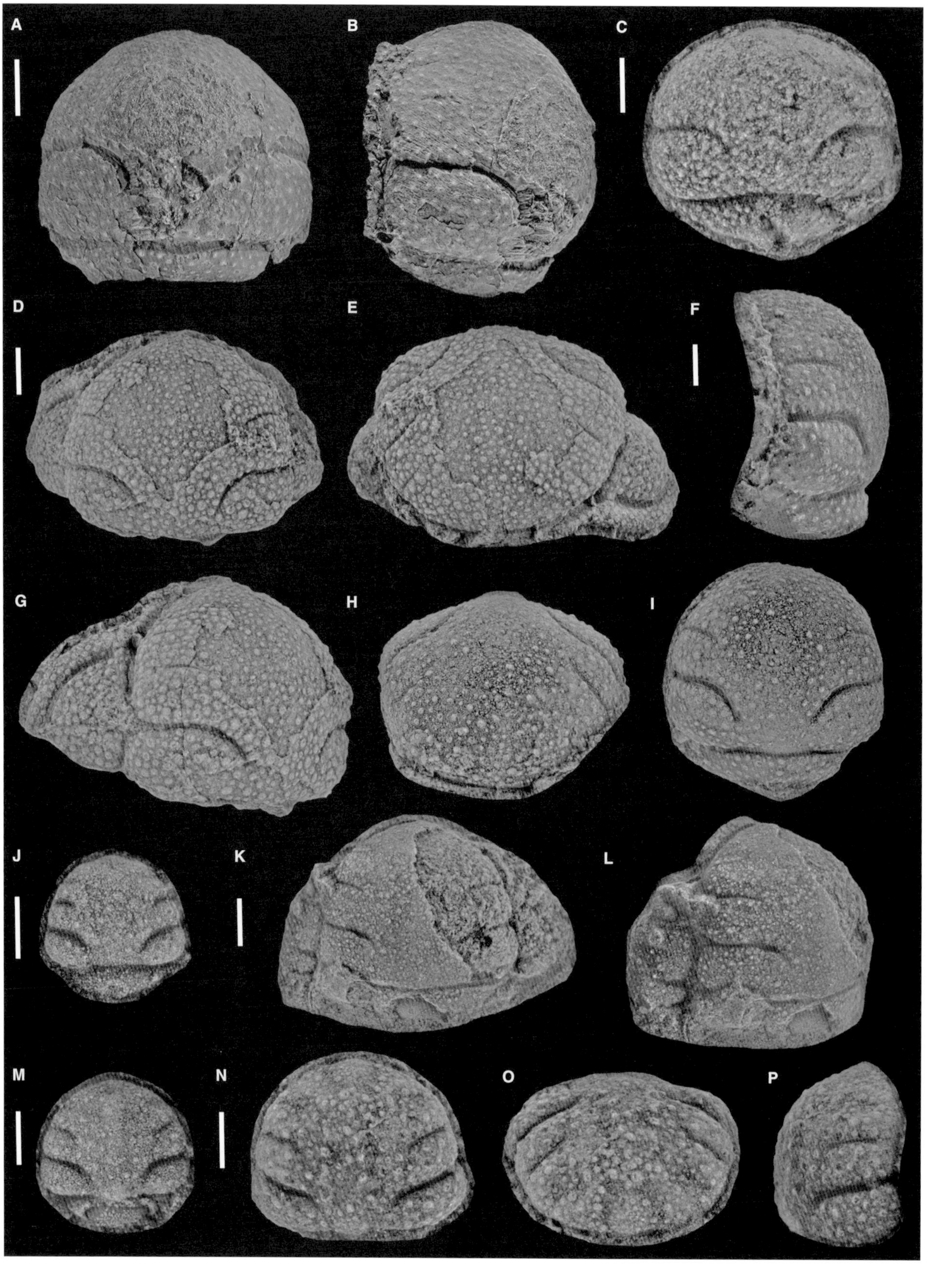
A
B
C
D
E
F
G
H
I
J
K
L
M
N
O
P

Diagnosis. See Edgecombe and Chatterton (1992, p. 53).

Distyrax bibullatus sp. nov.
Fig. 30A–E

LSID. urn:lsid:zoobank.org:act:9C2774C5-F250-4781-8F12-DD464875D724

Derivation of name. Combination of Latin, *bi*, two, double, and Latin, *bulla*, bubble, knob, boss, alluding to the tubercles present abaxially along posterior, as well as anterior, edge of occipital ring.

Holotype. MGUH30793 (Fig. 30A–C) cephalon, from GGU 275015 (KCL1).

Figured paratype. Locality KCL1: MGUH30794 pygidium.

Diagnosis. Tubercle present abaxially on posterior as well as anterior edge of occipital ring; sculpture of large tubercles and dense large pits on the pygidial pleural ribs.

Description. Highly convex cephalon measuring 7.5 mm long (sag.). Glabella excluding occipital ring roughly as long (sag.) as wide (tr.) across L4; narrowest across L1. Deep axial furrow. L1–L4 tuberculate; L1 smallest and subrounded, L2–L3 largest and elongate in direction of axial furrow. L4 larger than L1. Anterior to L4, axial furrow obscured, although PL is partially visible. Other glabellar tubercles comparable in size to that of L1, arranged irregularly, and with a narrow range of sizes. S1–S4 long (exsag.). Preglabellar furrow weak so anterior border of cranidium undifferentiated. Anterior border of cranidium and precranidial lobe with tubercles similar in size and distribution to those on frontal glabellar lobe. Length of palpebral lobe roughly 20% sagittal length of cephalon; midpoint opposite posterior of S3; raised, with tubercles aligned on raised area of palpebral lobe. Area of fixigena immediately adjacent to palpebral furrow lacking tubercles. Fixigenal tubercles overhanging axial furrow larger than those on rest of fixigena, but smaller than lateral lobe tubercles; other fixigenal tubercles of similar size and density to those on glabella. Posterior border furrow deep and long (exsag.); posterior border increasing in length (exsag.) abaxially, bearing a single row of tubercles. Genal spine present, but broken near base. Lateral border narrower (tr.) than librigenal field, but at least twice as wide (tr.) as posterior border is long (exsag.) and slightly increasing in width (tr.) anteriorly. Two or three rows of tubercles on lateral border (excluding those overhanging abax-

ial margin). Lateral border furrow shallower than posterior border furrow.

Pygidium fragmentary, discernible characters are as follows: high convexity; posterior of axis with ring furrows separated from axial furrow and absent medially; median tubercle present on every fourth ring; interpleural furrows deep and wider abaxially; pleural ribs strongly pitted except distally where they are upturned slightly; up to five randomly placed tubercles on each pleural rib, comparable in size to those on axis.

Remarks. Both the cephalic tuberculation patterns and pygidial morphology are characteristic of the genus. The posterior abaxial tubercle on the occipital ring of *D. bibullatus*, and the sculpture comprising a combination of large tubercles and dense large pits on the pygidial pleural ribs, are unique to the genus. Although the available material is limited, it is well preserved and cannot be assigned to any previously named species of *Distyrax*. Both cephalon and pygidium most resemble the type species, *D. peeli*, also from North Greenland. Notable differences are that *D. bibullatus* has a weak preglabellar furrow so that the anterior border of the cranidium is undifferentiated, its palpebral lobe is more anteriorly positioned, and the cephalic tubercles are smaller and more numerous. The cephalic morphology of *D. bibullatus* additionally resembles *D. cooperi* Edgecombe and Chatterton, 1992, particularly with respect to the complexity of the glabellar tuberculation, the shallow preglabellar furrow and placement of the eye. Edgecombe and Chatterton (1992, p. 59, table 1, and fig. 6) conducted a cladistic analysis on the type species of *Distyrax*, *D. peeli*, and their newly described species from the Llandovery of the Mackenzie Mountains, Canada (including *D. cooperi*), using *Encrinurus* as the outgroup. When available cephalic characters are coded for *D. bibullatus*, this species exhibits no plesiomorphic character states, and has most synapomorphies with *D. cooperi*.

Occurrence and distribution. The cephalon and pygidium are both from Kronpins Christian Land (KCL1).

Distyrax? sp.
Figure 30F

Remarks. A fragmentary cranidium from Central Peary Land, with a sagittal length of 8.2 mm, is preserved as an

---

FIG. 29. A–C, *Hyrokybe pharanx*? Lane, 1972; A–B, MGUH30785 cranidium, GGU 274689 (CPL5); A, dorsal and B, lateral views; C, MGUH30786 cranidium, GGU 198137 (CPL2) dorsal view. D–J, M, *Hyrokybe*? sp.; D–E, G, MGUH30787 cephalon, GGU 274654 (CPL4); D, dorsal, E, anterior and G, oblique lateral views; F, H–I, MGUH30788 cranidium, GGU 274689 (CPL5); F, lateral, H, anterior dorsal, and I, dorsal views; J, MGUH30789 cranidium, GGU 274654 (CPL4) dorsal view; M, MGUH30790 cranidium, GGU 274654 (CPL4) dorsal view. K–L, Acanthoparyphinae gen. et sp. indet. 1; MGUH30791 cephalon, GGU 198226 (CPL1); K, dorsal and L, oblique lateral views. N–P, Acanthoparyphinae gen. et sp. indet. 2; MGUH30792 cranidium, GGU 274689 (CPL5); N, dorsal, O, anterior and P, oblique views. Scale bars represent 5 mm (A–B) and 1 mm (C–P).

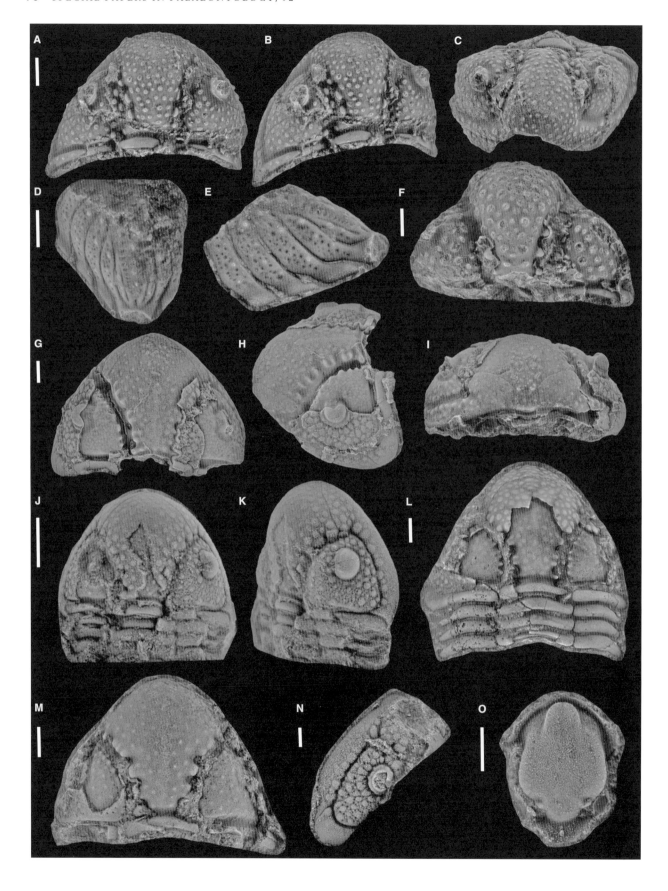

internal mould and has a similar glabellar morphology to *Distyrax bibullatus*. It is also highly convex, and the eye occupies a similar position, but the sculpture is much more coarsely tuberculate. As no occipital ring or librigena is preserved, generic placement is tentative.

Occurrence and distribution. Central Peary Land (CPL5).

Genus PERRYUS Gass and Mikulic, 1982

Type species. By original designation; *Perryus severnensis* Gass and Mikulic, 1982, from the Telychian – ?lower Wenlock Attawapiskat Formation, northern Ontario, Canada.

Other species. *Perryus bartletti* Edgecombe and Chatterton, 1992; *P. globosus* (Maksimova, 1962); *P. mikulici* sp. nov.; *P. palasso* (Lane, 1988).

Diagnosis. Modified after Gass and Mikulic (1982, p. 591) to incorporate observations by Edgecombe and Chatterton (1992, pp. 68–70), and other characteristics discussed below: glabellar furrow concave outward; L1 tuberculate abaxially; L2 and L3 comprising two partially fused longitudinally directed tubercles; midpoint of palpebral lobe opposite L3 or S3; librigenal border wide. Hypostome with anteriorly subquadrate rhynchos. Pygidium typically with 10–11 ribs and with axis typically comprising 20 or more rings.

Remarks. The characters of species of *Perryus* described here, and those of *P. bartletti* Edgecombe and Chatterton, 1992, and *P. palasso* (Lane, 1988), require the original concept of the genus to be modified, and the original diagnosis of Gass and Mikulic (1982) is revised to reflect this. The concave-outward glabellar furrow and partially fused nature of the tubercles of L2–L3 are particularly distinctive synapomorphies. In their discussion of *Cromus* Barrande, 1852, Edgecombe and Chatterton (1992, p. 62) removed plesiomorphic characters such as the presence of a preglabellar furrow and indistinct longitudinal median glabellar furrow, the complex glabellar tubercle array, the subquadrate to oval lateral glabellar lobe morphology, and a rounded genal angle from Strusz's (1980) diagnosis. Such characters are

similarly removed from Gass and Mikulic's (1982) diagnosis of *Perryus*. Additionally, the presence of transversely elongated alternate tubercles on the lateral border is specific to *P. severnensis*, and the general presence of tubercles on the lateral border of librigena is not unique to *Perryus*; these characters are removed also. There is a greater variation in the number of pygidial axial rings and pleural ribs, and pygidial sculpture, than documented by Gass and Mikulic (1982). Edgecombe and Chatterton (1992) suggested that the broad librigenal border may be diagnostic of *Perryus*, and this is supported by its presence in *P. mikulici* sp. nov. The subquadrate rhynchos of the hypostome of *P. palasso* and *P. bartletti* noted by Edgecombe and Chatterton (1992) is shared with *P. mikulici*.

Perryus mikulici sp. nov.
Figures 30G–O, 31A–D

LSID. urn:lsid:zoobank.org:act:1176F824-5F06-4F5E-860A-E90F1B726444

Derivation of name. For Dr D. G. Mikulic, geologist at the Illinois Geological Survey, who was an excellent guide for HEH to the Silurian of the Great Lakes area, USA.

Holotype. MGUH30796 (Fig. 30G–I) cephalon, from GGU 275015 (KCL1).

Figured paratypes. Locality CPL5: MGUH30797 cephalon with articulated thoracic segments; MGUH30799, 30801 cranidia; MGUH30800 librigena; MGUH30802 pygidium. Locality KCL1: MGUH30798 cranidium with articulated thoracic segments. Locality KCL6: MGUH30803 pygidium.

Diagnosis. L4 comprising three partially fused longitudinally directed tubercles, and PL a singular longitudinally directed tubercle; anterior of frontal glabellar lobe subtriangulate; glabellar tubercles markedly flattened, especially medially; tubercles non-perforate; pygidial axis with 20–22 rings some bearing subdued tubercles medially, and pleural regions with 11 pairs of ribs.

Description. Cephala and cranidia ranging in sagittal length from 4.9 to 13.3 mm; cephalic sagittal length two-thirds maxi-

---

FIG. 30. A–E, *Distyrax bibullatus* sp. nov.; A–C, MGUH30793 holotype cephalon; A, dorsal, B, oblique and C, anterior views; D–E, MGUH30794 pygidium; D, dorsal and E, oblique lateral views. Both specimens from GGU 275015 (KCL1). F, *Distyrax*? sp. MGUH30795 cranidium, GGU 274689 (CPL5) dorsal view. G–O, *Perryus mikulici* sp. nov.; G–I, MGUH30796 holotype cephalon, GGU 275015 (KCL1); G, dorsal, H, oblique lateral and I, anterior views; J–K, MGUH30797 cephalon with articulated thoracic segments, GGU 274689 (CPL5); J, dorsal and K, oblique lateral views; L, MGUH30798 cranidium with articulated thoracic segments, GGU 275015 (KCL1) dorsal view; M, MGUH30799 cranidium, GGU 274689 (CPL5) dorsal view; N, MGUH30800 librigena, GGU 274689 (CPL5) dorsal view; O, MGUH30801 cranidium, GGU 274689 (CPL5) dorsal view. All scale bars represent 1 mm.

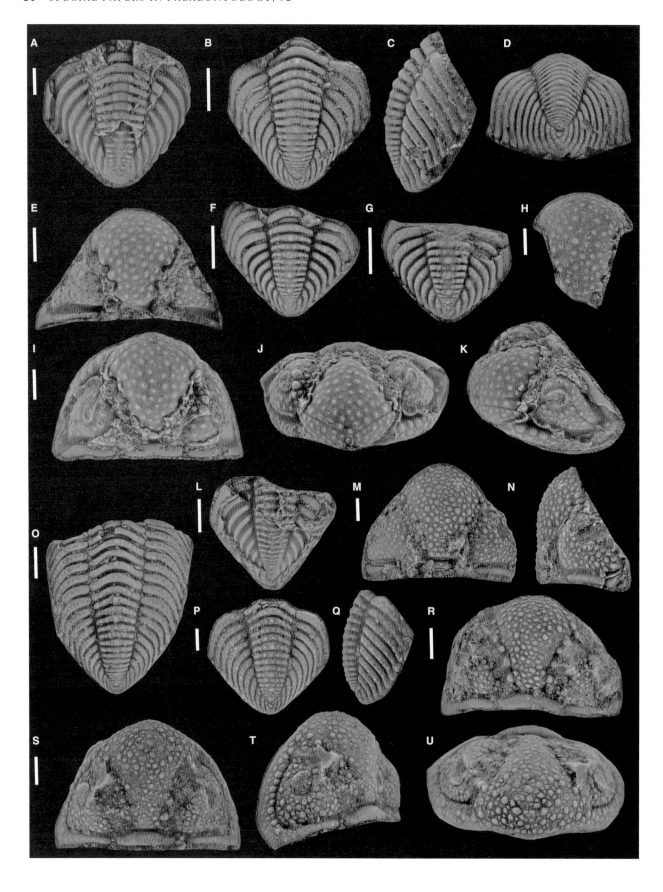

mum width (average 68%, range 60–78%; n = 7). Entire pre-occipital glabella, PL and precranidial lobe covered in unevenly sized and shaped, and irregularly distributed granulate tubercles. Tubercles are dorsally flattened, most so medially. Glabella, excluding occipital ring, roughly 90% as wide (tr.) across widest point – next to preglabellar furrow – as long (sag.). Axial furrows wide (tr.) and deep. L1 short (exsag.), with a singular transversely directed tubercle. Glabella wider at L1 than at L2; glabellar width (tr.) at L4 just over twice that across L2. S1–S3 narrow transversely, directed slightly antero-medially. PL with about 14 tubercles aligned alongside clear anterior border furrow, these increasing in size abaxially. Precranidial lobe becoming longer (exsag.) abaxially, with tubercles similar in size and distribution to those on frontal glabellar lobe. Occipital ring as wide as glabella at point near posterior of L4; 30% as long (sag.) as wide (tr.), two-thirds as long (exsag.) abaxially, as medially; flattened in lateral profile. Occipital furrow long (sag.) and deep. Palpebral lobe 15–20% total cranidial sagittal length; palpebral furrow well incised. Midpoint of palpebral lobe opposite L3; about three tubercles occur between palpebral lobe and axial furrow. Distance between palpebral lobes subequal to sagittal cranidial length. Genal field (excluding posterior border and palpebral lobe) covered in granulate tubercles, which are more even in size than those covering the glabella, and less flattened. Areas between tubercles distinctly pitted. Deep posterior border furrow. Posterior border of roughly equal length (exsag.) until abaxial of an exsagittal line through palpebral lobe, where it expands towards rounded genal angle; granulose sculpture becoming faintly tuberculate near genal spine. Posterior branch of facial suture increasingly posteriorly directed across lateral border. Lateral border of constant transverse width, wider (tr.) than librigenal field and posterior border length (exsag.). At least eight large tubercles alongside deep lateral border furrow, and less distinct tubercles abaxial of these. Connective sutures subparallel to sagittal line.

Hypostome at least 80% as wide (tr.) across anterior wings as long (sag.). Anterior wings located roughly one-quarter total hypostomal sagittal length back from anterior. Middle body with maximum width (tr.) in anterior lobe, one-third total sagittal length from posterior of middle body (just anterior of maculae): there measuring 73% sagittal length of middle body. Maculae obliquely elongate, located about 15% total sagittal length of middle body from posterior border of middle body. Rhynchos subquadrate, not projecting past hypostomal anterior border. Lateral margins of rhynchos gradually diverging and fading posteriorly, terminating completely by

one-quarter sagittal length of middle body from anterior margin. Anterior border subtriangulate; shorter (sag.) than anterior border furrow. Posterior hypostomal margin incomplete, but posterior border forming at least 13% entire hypostome sagittal length. Sculpture of fine granules covering entire middle body and posterior border.

Anterior of thorax with axis forming about 40% total transverse width. Thoracic segments without discernible sculpture.

Pygidial sagittal length ranging from 3.8 to 12.8 mm, averaging 97% as long as wide (range = 88–110%; n = 27). Axis comprising 41% total width of anterior of pygidium (range = 33–50%; n = 28). Axial rings medially with granulate sculpture and with irregular swellings, which in places are approaching tuberculate form. All rings longer sagittally than exsagittally, and inter-ring furrows deeper laterally, where they contain apodemal pits. Axial furrow well impressed. Pleural region with ribs displaying a granulate sculpture. Pleural ribs terminating as bluntly rounded expansions; these increasingly splayed outwards. Posteriormost three pairs merged posteroventrally. Interpleural furrows deep.

Remarks. *Perryus mikulici* is close to the type species, *P. severnensis* but differs from it by having a longer (sag.) glabella and cranidium, relative to their respective widths (tr.), and the precranidial lobe is less nearly semicircular in outline. Additionally, the precranidial lobe is entirely circumscribed in *P. mikulici*, whereas it is only distinct at its lateral extremities in *P. severnensis*. Internal moulds of *P. mikulici* (Fig. 30G–I, M) show L4 to comprise three partially fused tubercles; the nature of L4 is difficult to ascertain in *P. severnensis* as it is best seen on internal moulds which are not adequately preserved in the Attawapiskat material. The pygidial inter-ring furrows of *P. mikulici* are medially deeper and longer (sag.) than in *P. severnensis*. There are also considerable differences in the nature of the sculpture. *P. mikulici* has a flatter lateral profile to its glabellar tubercles, and they become effaced medially. Conversely, the pygidium bears a more pronounced sculpture; it is tuberculate in places, whereas there are no tubercles present in *P. severnensis*. Perforations are only present between tubercles on the genae in *P. mikulici*, whereas in *P. severnensis* these cover the entire body, and the tubercles themselves contain perforations. The distinctive morphology of tubercles covered

FIG. 31. A–D, *Perryus mikulici* sp. nov.; A, MGUH30802 pygidium, GGU 274689 (CPL5) dorsal view; B–D MGUH30803 pygidium, GGU 274788 (KCL6); B, dorsal, C, lateral and D, posterior views. E–G, *Perryus* cf. *P. palasso* Lane, 1988; E, MGUH30804 cranidium, GGU 298924 (SNL2) dorsal view; F, MGUH30805 pygidium with articulated thoracic segments, GGU 298506 (WL) dorsal view; G, MGUH30806 pygidium, GGU 298506 (WL) dorsal view. H–L, *Perryus* sp. 1; H, MGUH30807 glabella, GGU 198221 (CPL2) dorsal view; I–K, MGUH30808 cephalon, GGU 198206 (CPL2); I, dorsal, J, anterior and K, oblique lateral views; L, MGUH30809 pygidium, GGU 198214 (CPL2) dorsal view. M–Q, *Perryus* sp. 2; M–N, MGUH30810 cranidium; M, dorsal and N, lateral views; O, MGUH30811 articulated pygidium and thorax, dorsal view; P–Q, MGUH30812 pygidium, P, dorsal and Q, lateral views; all material from GGU 274774 (KCL5). R–U, genus aff. *Perryus*; R, MGUH30813 cranidium, dorsal view; S–U, MGUH30814 cephalon, S, dorsal, T, oblique lateral and U, anterior views; both specimens from GGU 198225 (CPL3). All scale bars represent 1 mm.

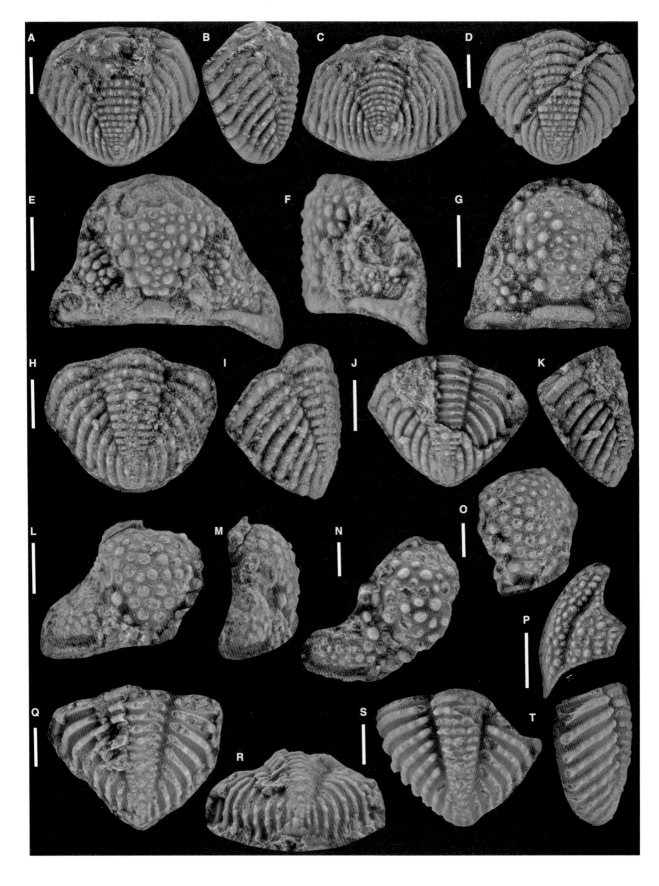

with granules is shared with both *P. severnensis* and *P. palasso*.

Occurrence and distribution. Most common in Kronprins Christian Land (KCL1, 6) and Central Peary Land (CPL5). Also found in Western Peary Land (WPL1).

Perryus cf. *P. palasso* (Lane, 1988)
Figure 31E–G

Remarks. Cranidia referable to *Perryus* from Wulff Land and South Nares Land bear short genal spines. This, and the much reduced L1, are characters of *P. palasso* from Peary Land (see Lane, 1988, pl. 2, figs 1–8, 10–13; pl. 3, figs 5–6, 11). The figured cranidium is almost twice as wide (tr.) as long (sag.), and the glabella excluding occipital ring is 92% as wide as long. The pygidia, including a pygidium with at least three articulated thoracic segments (Fig. 31F), are tentatively assigned to the same species as the cranidia on the basis of their co-occurrence. Nine pleural ribs are present and the pygidium is slightly wider than long, as in *P. palasso*. They differ from other *Perryus* species, including *P. palasso*, in having about 16 pygidial axial rings: the genus typically has 18–22. Unlike *P. palasso*, there are rows of tubercles on both axial rings and pleural ribs, and larger tubercles on five of the axial rings.

Occurrence and distribution. Found in Wulff Land (WL) and South Nares Land (SNL2).

Perryus sp. 1
Figure 31H–L

Remarks. A glabella, cephalon and pygidium are preserved as internal moulds with little cuticle attached, and as such only limited description is possible. L2 and L3 comprise pairs of partially fused and longitudinally directed tubercles which are characteristic of the genus. Tubercles where present exhibit the granulose form exhibited by most species of *Perryus*. Proportionally, the cephalon is similar to *P. mikulici* but has a distinct preglabellar furrow and inflated precranidial lobe, and tubercles are clear on the internal mould, which is not the case with

P. mikulici. Pygidia are relatively wider than those of *P. mikulici*, and comprise nine pleural ribs and 18–20 axial rings, fewer than in *P. mikulici*. A new species is represented, but additional material is required to name it.

Occurrence and distribution. Central Peary Land (CPL2).

Perryus sp. 2
Figure 31M–Q

Remarks. A cranidium has more strongly raised tubercles than those of *P. mikulici*. The associated pygidium and pygidium with articulated thorax exhibit additional differences: they have fewer axial rings (18 compared with 20–22), and rows of clear tubercles are present on both axial rings and pleural ribs.

Occurrence and distribution. Kronprins Christian Land (KCL5).

Genus aff. *Perryus*
Figures 31R–U, 32A–D

Remarks. Cephala and cranidia ranging in sagittal length from 3.1 to 4.9 mm have similar proportions to *P. mikulici*. They share the following similarities: asymmetrically arranged and granulate tubercles; the position of palpebral lobe opposite L3; lateral border wider (tr.) than librigenal field; pitted cheeks. The morphology of the lateral glabellar lobes is quite different from *Perryus* however, comprising individual subrounded tubercles as opposed to the diagnostic paired longitudinally-fused type.

Pygidia ranging in sagittal length from 4.3 to 4.9 mm are proportionally wider than those of *P. mikulici*, being 80% as long (sag.) as wide (tr.). The axis comprises a similar proportion of the total anterior width of the pygidium as in *P. mikulici*. There are fewer axial rings than any species of *Perryus* (only 15 are clearly discernable, compared with the 20+ rings characteristic of the genus). As with *P. mikulici*, the posteriormost pleural ribs are posteroventrally merged. The number of pleural ribs (10) is like *Perryus*.

FIG. 32. A–D, genus aff. *Perryus*; A–C, MGUH30815 pygidium; A, dorsal, B, lateral and C, posterior views; D, MGUH30816 pygidium, dorsal view; both specimens from GGU 198225 (CPL3). E–K, *variolaris* plexus gen. indet. 1; E–F, MGUH30817 cranidium, GGU 274689 (CPL5); E, dorsal and F, oblique lateral views; G, MGUH30818 cranidium, GGU 274996 (KCL4) dorsal view; H–I, MGUH30819 pygidium, GGU 274689 (CPL5); H, dorsal and I, oblique views; J–K, MGUH30820 pygidium, GGU 274689 (CPL5); J, dorsal and K, oblique views. L–T, *variolaris* plexus gen. indet. 2; L–M, MGUH30821 cranidium; L, dorsal and M, oblique views; N, MGUH30822 cranidium, dorsal view; O, MGUH30823 glabella, dorsal view; P, MGUH30824 librigena, dorsal view; Q–R, MGUH30825 pygidium; Q, dorsal and R, posterior views; S–T, MGUH30826 pygidium; S, dorsal and T, lateral views; all material from GGU 198226 (CPL1). All scale bars represent 1 mm.

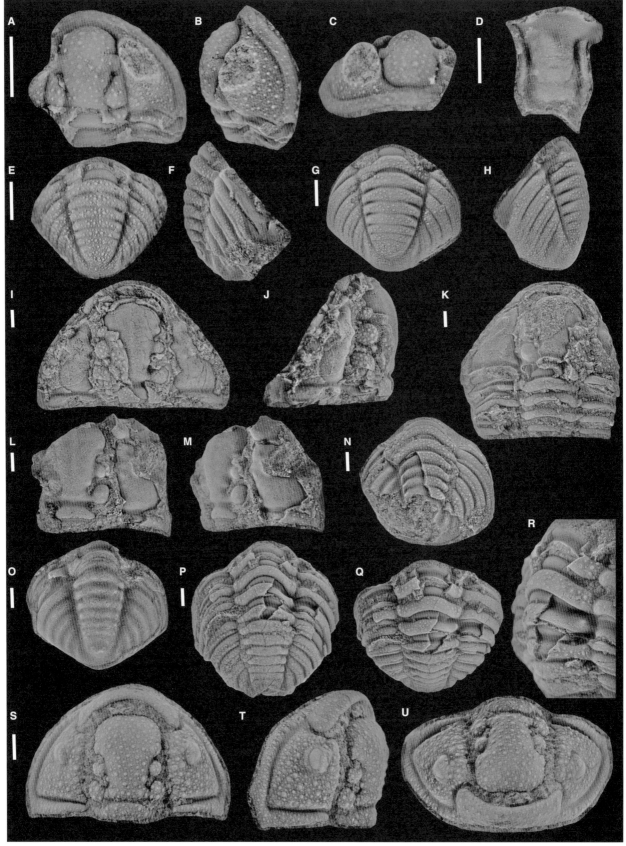

Occurrence and distribution. All material is from Central Peary Land (CPL5).

variolaris plexus gen. indet. 1
Figure 32E–K

Remarks. Generic assignment of these cranidia and pygidia is difficult based on such limited material. The fragments show characters consistent with those of the '*Encrinurus*' *variolaris* plexus of Strusz (1980): enlarged tuberculiform lateral lobes anterior of L1; L1 reduced; complex glabellar tuberculation (tuberculation includes I-1, small ii-0, II-1 abaxial to I-1, iii-0, III-1,2 (III-2 partially merged with L3), IV-1,2,3 (IV-3 partially merged with L4) and much smaller, irregularly distributed tubercles present between these); genal spines small and thorn-like; subtriangular pygidium; tuberculate pleurae, last one or two recurved, loop-like; pygidium non-mucronate and lacking pygidial spine. The *variolaris* plexus has been well studied, including cladistically (Adrain and Edgecombe 1995; Edgecombe and Ramsköld 1996), and differentiation made between typical Llandovery taxa such as *Nucleurus* Ramsköld, 1986 and *Billevittia* Edgecombe and Ramsköld, 1996, and post-Llandovery clades which include *Balizoma* Holloway, 1980 and *Struszia* Edgecombe, Ramsköld and Chatterton *in* Edgecombe and Chatterton, 1993. The material has characters in common with both Llandovery and post-Llandovery genera, and it is left under open nomenclature.

Occurrence and distribution. Central Peary Land (CPL5) and Kronprins Christian Land (KCL4).

variolaris plexus gen. indet. 2
Figure 32L–T

Remarks. Fragmentary cranidia, pygidia and free cheeks from Central Peary Land are all preserved as internal moulds. The cranidia are reminiscent of *variolaris* plexus gen. indet. 1 but have larger, more evenly sized glabellar tubercles. The pygidia are similarly proportioned too (about 80% as long (sag.) as wide with axis comprising

about one-third pygidial sagittal width across widest point) and have the same number of axial rings (about thirteen) and pleural ribs (eight or nine).

Occurrence and distribution. All material is from Central Peary Land (CPL1).

Suborder CALYMENINA Swinnerton, 1915
Family CALYMENIDAE Burmeister, 1843
Subfamily CALYMENINAE Burmeister, 1843

Genus CALYMENE Brongniart *in* Brongniart and Desmarest, 1822

Type species. *Calymene blumenbachii* Brongniart *in* Desmarest, 1817, from the Wenlock (Homerian Stage) Much Wenlock Limestone Formation, Dudley, West Midlands, England, by subsequent designation of Shirley (1933, p. 53).

Diagnosis. See Siveter (1985).

Remarks. A great many species have been assigned to this genus, sometimes rather indiscriminately, and no comprehensive phylogenetic analysis has been undertaken. Pending a thorough revision of the genus, no list of other species is presented. Edgecombe and Adrain (1995) included *C. brownsportensis* Edgecombe and Adrain, 1995, *C. clavicula* Campbell, 1967 and *C. resticta* Prouty, 1923, in *Calymene* s.s., a core species-group most closely related to *C. blumenbachii*. Siveter (1996) revised species of *Calymene* from the Wenlock of the UK, including additional species which undoubtedly belong to *Calymene* s.s.

Calymene aff. iladon Lane and Siveter, 1991
Figure 33A–H

Remarks. The nature of the anterior border, and the connection of L2 to the fixigena, in a cephalon measuring 2.3 mm (sag. length) indicate the generic assignment. The cephalon is closest to *C. iladon* and shares the following distinctive characters with that species: glabella weakly bell-shaped (83% as wide (tr.) across L1, as long (sag.));

FIG. 33. A–H, *Calymene* aff. *iladon* Lane and Siveter, 1991; A–C, MGUH30827 cephalon; A, dorsal, B, lateral and C, anterior views; D, MGUH30828 hypostome, ventral view; E–F, MGUH30829 pygidium; E, dorsal and F, lateral views; G–H, MGUH30830 pygidium; G, dorsal and H, lateral views; all material from GGU 298506 (WL). I–R, *Calymene* sp.; I–J, MGUH30831 cranidium, GGU 198185 (CPL2); I, dorsal and J, lateral views; K, MGUH30832 cranidium with articulated thoracic segments, GGU 198223 (CPL2) dorsal view; L–M, MGUH30833 cranidium, GGU 198223 (CPL2); L, dorsal and M, oblique views; N, MGUH30834 pygidium, GGU 198196 (CPL2) dorsal view; O, MGUH30835 pygidium, GGU 198214 (CPL2) dorsal view; P–R, MGUH30836 pygidium with articulated thoracic segments, GGU 198223 (CPL2); P, dorsal view of pygidium, Q, dorsal view of thoracic segments and R, oblique lateral view of thoracic segments. S–U, Calymenina indet. MGUH30837 cephalon, GGU 301318 (WPL1); S, dorsal, T, oblique lateral and U, anterior views. All scale bars represent 1 mm.

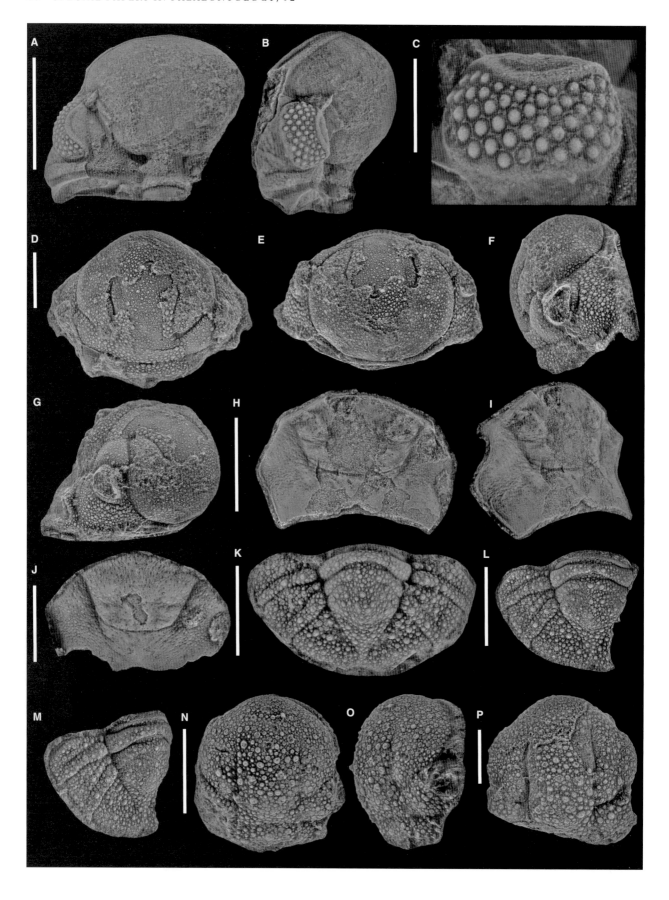

relatively short preglabellar area (of subequal length (sag.) to anterior border); forward positioning of the palpebral lobe; complete covering of evenly spaced coarse granules. Some differences from *C. iladon* are seen, but these could reflect the small size of the cephalon, at least in part: fork of S1 with less distinctive intermediate lobe; L1 lobe distinctly more triangulate in form; weak L2-fixigena buttress. An associated hypostome demonstrates the extremely weakly raised macula and rhyncos, as seen in *C. iladon*. The pygidium is associated on the basis of its co-occurrence with the cephalon, but it is larger in size, measuring 3.8 mm long (sag.). The wide (tr.) axis, comprising roughly half the width (tr.) of the pygidium anteriorly, distinguishes the pygidium from that of *C.* sp. 1, and it is also proportionally wider than that of *C. iladon*. The number of axial rings and pleural ribs, and the granular sculpture do agree with *C. iladon*, however.

The coarse tuberculate sculpture seen in *C. iladon* and *C.* aff. *iladon* is unusual among the calymenines; it is a characteristic shared with *C.* (s.l.) aff. *iladon* (of Adrain *et al.* 1995), and *Arcticalymene* Adrain and Edgecombe, 1997, from the Wenlock Cape Phillips Formation, Central Canadian Arctic.

Occurrence and distribution. All material is from Wulff Land (WL).

Calymene sp.
Figure 33I–R

Remarks. Description of *C.* sp. is constrained by the limited material, and the patchy nature of cuticle cover, however, the papillate L2 opposite a distinct buttress on the fixigena is generically diagnostic. *C.* sp. differs from other described *Calymene* species from the Silurian of North Greenland, of which *C. iladon* Lane and Siveter, 1991 from the Telychian Offley Island Formation is the most complete known. *C.* sp. has the midpoint of the palpebral lobe placed opposite the midpoint of L2, whereas the palpebral lobe of *C. iladon* is very forwardly placed. The species have a very similar sculpture of variably sized and spaced granules.

Calymene sp. of Lane (1972, p. 359, pl. 64, figs 6a–d, 7a–b) from the Telychian Samuelsen Høj Formation of Kronprins Christian Land, eastern North Greenland, has a more convex (sag.) glabella, with the frontal lobe almost overhanging the anterior border. That material was reassigned to *Arcticalymene* by Adrain and Edgecombe (1997). A fragmentary cranidium from south-eastern Hall Land on the northern slopes of Kayser Bjerg (Telychian Offley Island and Hauge Bjerge formations) (Lane 1984, p. 62, pl. 2, fig. 10) is thought to be conspecific with Lane's 1972 material (Lane and Siveter, 1991).

Occurrence and distribution. From Central Peary Land (CPL1–2).

Calymenina indet.
Figure 33S–U

Remarks. A cephalon measuring 64% as long (sag.) as wide (tr.) is referred to Calymenina indet. It lacks papillae on the fixigena opposite L2, a characteristic of *Flexicalymene* and *Gravicalymene*, and unlike *Calymene*, *Arcticalymene* and *Diacalymene*, but exhibits other characteristics more like some of the latter genera. These include the anterior position of the eye and the sculpture of variably sized and spaced granules like *Arcticalymene*. The cephalon is particularly distinctive with respect to the exsagittal deflection displayed by the anterior border furrow and the subrectangular glabellar morphology.

Occurrence and distribution. The cephalon is from Western Peary Land (WPL1).

Suborder PHACOPINA Struve *in* Moore, 1959
Family PHACOPIDAE Hawle and Corda, 1847
Subfamily PHACOPINAE Hawle and Corda, 1847

Genus ACERNASPIS Campbell, 1967

Type species. By original designation: *Phacops orestes* Billings, 1860, from the Llandovery Jupiter and Gun River Formations, Anticosti Island, Canada.

Diagnosis. See Campbell (1967, p. 32), Ramsköld (1988), and Ramsköld and Werdelin (1991).

Remarks. Campbell (1967) erected the genera *Acernaspis* and *Ananaspis*. Ramsköld (1988) described the early growth stages of *Acernaspis* and found similarities between its ontogenetic stages and the genus *Ananaspis*,

FIG. 34. A–C, *Acernaspis*? sp. MGUH30838 cephalon, GGU 298506 (WL); A, dorsal view, B, oblique lateral view and C, extended view of visual surface. D–M, *Dicranogmus pearyi* sp. nov.; D–G, MGUH30839 holotype cephalon; D, dorsal, E, anterior, F, lateral and G, oblique views; H–I, MGUH30840, hypostome; H, ventral and I, oblique lateral views; J, MGUH30841, hypostome, ventral view; K, MGUH30842 pygidium, dorsal view; L–M, MGUH30843 pygidium; L, dorsal and M, oblique lateral views; all material from GGU 274689 (CPL5). N–P, *Dicranogmus* sp.; N–O, MGUH30844 cranidium, GGU 274654 (CPL4); N, dorsal and O, lateral views; P, MGUH30845 cranidium, GGU 198226 (CPL1) dorsal view. All scale bars represent 5 mm.

which he interpreted as arising through neoteny. Ramsköld and Werdelin (1991) conducted a cladistic analysis on a number of phacopid trilobites and redefined *Acernaspis* and *Ananaspis* to make them monophyletic.

Acernaspis? sp.
Figure 34A–C

Remarks. A single, fragmentary cephalon has a sagittal length of 7.8 mm. The glabella, excluding the occipital ring, is 87% as long (sag.) as wide (tr.) across the frontal lobe, and the anterior margin is smoothly rounded. The preglabellar furrow is clearly impressed both abaxially and medially. The palpebral lobe has a shallow palpebral furrow and comprises about one-third total cephalic sagittal length, having its posterior margin placed just anterior of L1. The eye is well preserved; the visual surface has sixteen files, with a lens formula: 234 34**4** **4**44 434 343 2; total 55 (Fig. 34C; lenses in bold are significantly smaller than others). The glabellar morphology and position of the palpebral lobe suggest that this cephalon is closest to *Acernaspis*. The glabella is much more highly inflated in lateral profile than in other *Acernaspis* species, however, and the cephalon could represent an undescribed related genus.

Occurrence and distribution. Found in Wulff Land (WL).

Order LICHIDA Moore, 1959
Family LICHIDAE Hawle and Corda, 1847
Subfamily TROCHURINAE Phleger, 1936

Genus DICRANOGMUS Hawle and Corda, 1847

Type species. By monotypy; *Dicranogmus pustulatus* Hawle and Corda, 1847, from the Ludlow Kopanina Formation, Czech Republic.

Other species. Dicranogmus aequalis (Törnquist, 1884); *D. confinis* Apollonov, 1980; *D. gregalis* Chugaeva, 1983; *D. guizhouensis* Wu, 1977; *D. lepidus* Campbell and Chatterton, 2006; *D. pearyi* sp. nov.; *D. scabrosus* Zhou and Zhou, 1982; *D. simplex* (Barrande, 1846); *D. skinneri* Perry and Chatterton, 1977; *D. wilsoni* Campbell and Chatterton, 2006; *D. wynni* Adrain, 2003.

Diagnosis. Cranidial features from Thomas and Holloway (1988, p. 230), and librigenal and pygidial features from Adrain (2003, p. 751), here modified to reflect the partial definition of the second and third pygidial inter-ring furrows seen in some species: trochurine with strongly convex (sag., exsag.) glabella overhanging anterior border; in dorsal view, glabella wider across bullar lobes than long (sag.) and smoothly curved in outline anteriorly, median and bullar lobes there lacking independent convexity.

Longitudinal furrows subparallel in front of S1, usually dying out approximately halfway to front of bullar lobe, may extend behind S1 as poorly defined depressions not interrupting exoskeletal granulation. Maximum width of bullar lobe almost equal to width of median lobe. L1a usually clearly circumscribed; S1 may be present behind median lobe as weak, concave-forward depression not interrupting exoskeletal granulation and meeting occipital furrow on sagittal line; posterior and lateral border furrows shallow or effaced; genal spine broad and short. Cephalon with dense tuberculate sculpture; lateral border with dorsal row of large tubercles. Pygidium with anteriorly broad axis on which only the first inter-ring furrow is fully defined, second and third inter-ring furrows may be apparent abaxially; strongly tapering, subtriangular postaxial band set off from axis by abrupt change in slope; three sets of subtriangular to pointed border spines; first and second pygidial pleural and interpleural furrows fully defined; third pleural and interpleural furrows defined only as minute proximal depressions.

Remarks. Dicranogmus? sp. of Lane and Owens, 1982, *D. pearyi* sp. nov. and *D.* sp. are the only *Dicranogmus* species from North Greenland with described cranidia, hypostomes and pygidia. They differ from the Wenlock Canadian Arctic species *D. skinneri* Perry and Chatterton, 1977 and *D. wynni* Adrain, 2003, and from the Wenlock north-western Canadian species *D. wilsoni* Campbell and Chatterton, 2006 and *D. lepidus* Campbell and Chatterton, 2006, in having: anteriorly reduced glabellar tubercles; better impressed hypostomal middle furrows; posterior border furrow curved posteriorly as opposed to running transversely with a slight anterior deflection; posterior border abaxially more posteriorly extended; second and third pygidial axial ring furrows are visible abaxially.

Dicranogmus pearyi sp. nov.
Figure 34D–M

LSID. urn:lsid:zoobank.org:act:4A8C7086-A349-476A-A50B-1C71AA35D3F4

Derivation of name. For Admiral Robert Edwin Peary, who carried out seven expeditions to North Greenland, reaching the North Pole in 1909, and for whom Peary Land, where this material was collected, was named.

Holotype. MGUH30839 (Fig. 34D–G) cephalon; from GGU 274689 (CPL5).

Figured paratypes. Locality CPL5: MGUH30840–30841, hypostomes, MGUH30842–30843 pygidia.

Diagnosis. Glabellar tubercles becoming finer anteriorly; longitudinal furrows deflected abaxially near anterior of level of palpebral lobe; faint posterior extension behind S1 running almost exsagittally; hypostomal lateral border furrow continually diverging anteriorly throughout its course; almost intersecting hypostomal anterior margin; second and third axial ring furrows visible abaxially; postaxial band strongly triangulate, roughly one-third total pygidial axis length (sag.).

Description. Cephalon with sculpture of unevenly sized and shaped tubercles. Glabella excluding occipital ring on average 83% as long (sag.) as wide (tr.) across bullar lobes (range = 80–87%; n = 2); front of glabella with slight concavity anterolaterally (Fig. 34E). Longitudinal furrow visible for posteriormost 64% of pre-occipital glabella (range = 63–66%; n = 2). Longitudinal furrow deep, terminating a very short distance anterior of palpebral lobe in lateral view. Median glabellar lobe wider (tr.) than bullar lobe is long (exag.). S1 deep, extending anterolaterally to meet glabellar furrow at position opposite anterior half of palpebral lobe. Occipital furrow deepest medially; occipital lobe effaced. Well defined preglabellar furrow and narrow (sag., exsag.) anterior border. Genal field highly convex longitudinally, with sculpture most comparable to that on occipital ring. Eye socle widening (tr.) anteriorly; lower margin particularly well incised. Lateral border furrow shallow.

Hypostome subtrapezoidal. Middle body averaging 65% as long (sag.) as wide (tr.) anteriorly (62–67%; n = 2), maximum width about 66% hypostomal width across shoulders. Middle furrow subtransverse, one-quarter to one-third width of middle body at this point. Lateral border increasing rapidly in width posteriorly, widest opposite a point between maculae and middle furrow. From there, lateral margin projected posteromedially. Posterior border almost equal in length (sag., exsag.) posterior to median body, slightly narrower medially. Abaxial of median body, posterior margin smoothly curving posteriorly so that posterior border is posterolaterally pointed. Sculpture comprising elongated pits and striations, sparsely tuberculate around anterolateral borders of middle body.

Pygidium semicircular; 54% as long (sag.) as wide (tr.) (n = 1). Axis comprising 40% (n = 1) pygidial width (tr.) anteriorly, gently narrowing posteriorly; first inter-ring furrow well impressed; second inter-ring furrow less well impressed, effaced in medial third; third inter-ring furrow visible abaxially for less than this. Terminal piece inflated; postaxial band with furrows converging and almost meeting, not intersecting posterior pygidial margin. Interpleural furrows deep, defining two clear marginal spines with posterior pleural bands inflated adjacent to axial furrow. Pygidial marginal spines with rounded anterior margins and spinose posterior margins. Pleural furrows shallowing close to margin. Sculpture of densely packed, unevenly sized tubercles.

Remarks. The high convexity glabella, anterior effacement of the longitudinal furrows and lack of a deep pleural furrow on the posteriormost pygidial pleural segment are particularly characteristic of the genus.

D. pearyi is not directly comparable with any other species of *Dicranogmus*, other than *D.* sp. described below. It is compared, however, with the only other species of *Dicranogmus* known in sufficient detail from North Greenland: *Dicranogmus*? sp. Lane and Owens, 1982, from blocks believed to have been derived from the Telychian Pentamerous Bjerg Formation, of Kap Schuchert, Washington Land, western North Greenland. Synapomorphies include: glabellar outline (although the anterior of the glabella in *D. pearyi* exhibits a slight anterolateral concavity); anterior deflection of longitudinal furrow; length (sag., exsag.) of both the anterior border and the occipital ring; convexity of genal field; eye socle with lower margin well incised, widening anteriorly; position and relative size of palpebral lobe; pygidial and pygidial axis proportions; morphology of the pygidial axis and pleural and interpleural furrows. The distinguishing characters of the dorsal surface are the larger and more unevenly sized tubercles, and the effaced glabellar furrows in *D.*? sp. The hypostomal morphology exhibits the following differences: sculpture with elongated pits and striations, and only sparsely tuberculate in *D. pearyi*, whereas that of *D.*? sp. is more tuberculate, especially on the lateral border; posterior margin smoothly curved in *D. pearyi*, compared with a clear indentation in the posterior margin of *D*? sp. which exhibits a pair of posterolateral projections.

Occurrence and distribution. Found in Central Peary Land (predominantly CPL5, also CPL4).

Dicranogmus sp.
Figures 34N–P, 35A–E

Description and remarks. As for *D. pearyi* except that: glabella excluding occipital ring relatively longer, on average 91% as long (sag.) as wide (tr.) across bullar lobes (range = 87–96%; n = 6); longitudinal furrows subparallel and persistent, lacking the abaxial inflection seen in *D. pearyi*, and visible for posteriormost 72% of pre-occipital glabella (range = 61–87%; n = 6), so less effaced anteriorly; three shallow furrows radiating posteriorly from confluence of S1 and longitudinal furrows; coarser tubercles; hypostome with tuberculate sculpture, hypostomal lateral border furrow anteriorly curved adaxially, terminating prior to hypostomal anterior margin, so that short (exsag.) anterior border present abaxially; middle furrow running slightly posteromedially; posterior margin more strongly indented; sculpture tuberculate rather than striated and pitted; pygidial axis relatively narrower (tr.); terminal piece less inflated in lateral profile; postaxial band comprising a much greater sagittal length of pygidium, only defined anteriorly; both first and second pleural furrows terminating abruptly roughly 15% total spine length (oblique tr.) from pygidial margin; sculpture more finely tuberculate.

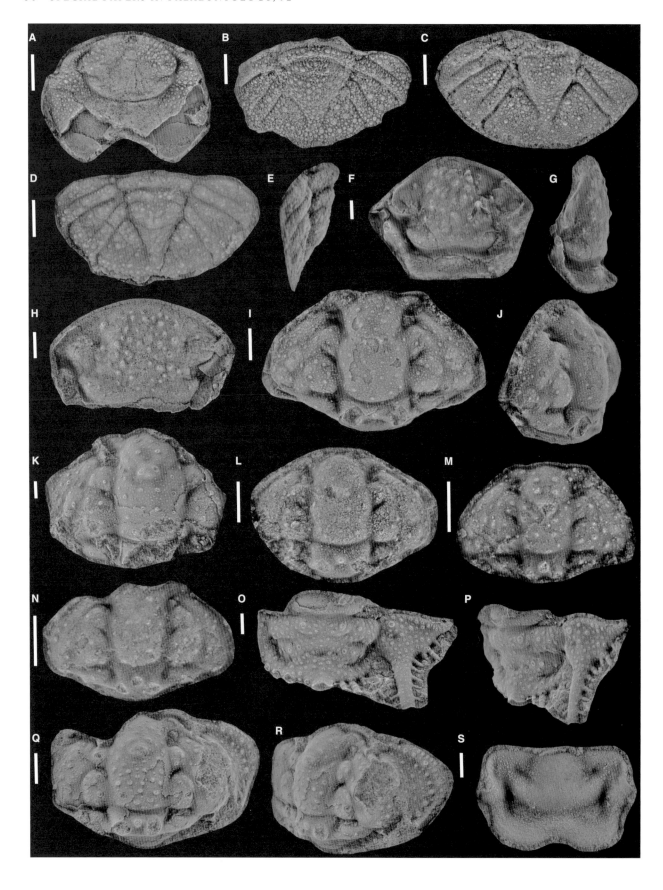

Occurrence and distribution. Central Peary Land (predominantly CPL1, also CPL2, 4).

Trochurinae indet.
Figure 35F–H

Remarks. Two hypostomes with middle bodies a little wider (tr.) across maculae than long (sag.), have the characteristic proportions and effaced median furrow of trochurines. The very coarse, irregularly shaped and laterally elongated tubercles on the anterior lobe of the middle body are particularly reminiscent of *Acanthopyge*. The hypostomes are not determinable generically and may not belong to a single species.

Occurrence and distribution. The two hypostomes are from Central Peary Land (CPL5).

Order ODONTOPLEURIDA Whittington *in* Moore, 1959
Family ODONTOPLEURIDAE Richter and Richter, 1917
Subfamily CERATOCEPHALINAE Richter and Richter, 1925, as emended by Prantl and Přibyl (1949)

Remarks. Ramsköld (1991) revised parts of the Odontopleuridae using cladistic analyses and recognized the Subfamily Ceratocephalinae, comprising *Ceratocephala* Warder, 1838 and *Ceratocara* Ramsköld, 1991.

Genus CERATOCEPHALA Warder, 1838

Type species. By monotypy; *Ceratocephala goniata* Warder, 1838, from the Middle Silurian of Springfield, Ohio.

Other species. Ceratocephala angostura Šnajdr, 1986; *C. avalanchensis* Chatterton and Perry, 1983; *C. barrandii* (Fletcher *in* Salter, 1853); *C. biscuspis* (Angelin, 1854); *C. bowningensis* (Etheridge and Mitchell, 1917); *C. coalescens* Van Ingen, 1901; *C. confinis* Tripp, 1962; *C. confraga* Raymond, 1925; *C. crawfordi* Chatterton and Perry, 1983; *C. depauperata* Van Ingen, 1901; *C. exigua* Whittington, 1963; *C. graffhami* Shaw, 1974; *C. hoernesi* (Hawle and Corda, 1847); *C. horani* (Billings, 1857); *C. laciniata* Whittington and Evitt, 1954; *C. laticapitata* (Warburg, 1925); *C. lochkoviana* Chlupáč, 1971; *C. micula* Ancygin, 1977; *C. nipponica* Kobayashi and Hamada, 1977; *C. osklundi* Thorslund, 1940; *C. plummeri* Chatterton and Perry, 1983; *C. rara* (Barrande, 1872); *C. relativa* Tripp, 1967; *C. rhabdophora* (Hawle and Corda, 1847); *C. triancantheis* Whittington and Evitt, 1954; *C. tungstenensis* Chatterton and Perry, 1983; *C. verneuili* (Barrande, 1846); *C. vesiculosa* (Beyrich, 1846); *C. vexilla* Chatterton, 1971; *C. vogdesi* Etheridge and Mitchell, 1896; *C. vranovica* Šnajdr, 1986.

Diagnosis. See Thomas (1981, p. 93).

Ceratocephala sp. 1
Figure 35I–P

Remarks. Cranidia ranging in sagittal length from 2.1 to 6.8 mm and on average 65% as long (sag.) as wide (tr.) (range = 62–71%; n = 6) are assigned to *Ceratocephala* on the basis of their glabellar lobe morphology. The cranidia show three main differences and could represent different species of *Ceratocephala*. Alternatively, given the range in sclerite size, the differences could be ontogenetic in nature: smaller specimens (MGUH30855–30856; Fig. 35M–N) have increasingly distally fused L1 and L2, these becoming more abaxially distinct in larger specimens (MGUH30852–30853; Fig. 35I–K); the median glabellar lobe is increasingly swollen opposite L2 in larger specimens; in smaller specimens, the median spine base on the occipital ring is subequal in size to the ones abaxial to it; in larger specimens, the median spine base is reduced in size relative to those abaxially. The overall cranidial morphology is similar to that of *C. avalanchensis* Chatterton and Perry, 1983 (p. 48, pl. 31, figs 5–6, 9, 23–24; pl. 32, figs 7, 10, 15, 18–19, 21, 24; pl. 33, fig. 17; text-fig. 40) from the Llandovery Whittaker Formation, Mackenzie Mountains, Canada; however, the cranidium is distinguished from the Mackenzie Mountains species by the less inflated (tr.) median glabellar lobe and lack of an anterolateral swelling. These conditions are present in Ordovician species such as *C. laticapitata* Warburg, 1925 from the Boda Limestone Formation, Kallholn, Siljan District, redescribed and figured by Bruton (1966, p. 24, pl. 4, figs 5–8; pl. 5, figs 6–8), and also Silurian species such as *C.* cf *C. biscuspis* (Angelin, 1854) of Lane and Owens (1982), which is the only other *Ceratocephala* cephalon previously described from North Greenland. *C.* sp. 1 differs from *C. laticapitata*

FIG. 35. A–E, *Dicranogmus* sp.; A, MGUH30846 hypostome, GGU 198226 (CPL1) ventral view; B, MGUH30847 pygidium, GGU 198137 (CPL2) dorsal view; C, MGUH30848 pygidium, GGU 198226 (CPL1) dorsal view; D–E, MGUH30849 pygidium, GGU 198226 (CPL1); D, dorsal and E, lateral views. F–H, Trochurinae indet.; F–G, MGUH30850 hypostome; F, ventral and G, lateral views; H, MGUH30851 hypostome, ventral view; both specimens from GGU 274689 (CPL5). I–P, *Ceratocephala* sp. 1; I–J, MGUH30852 cranidium, GGU 274689 (CPL5); I, dorsal and J, oblique lateral views; K, MGUH30853 cranidium, GGU 274654 (CPL4) dorsal view; L, MGUH30854 cranidium, GGU 274689 (CPL5) dorsal view; M, MGUH30855 cranidium, GGU 275015 (KCL1) dorsal view; N, MGUH30856 cranidium, GGU 198226 (CPL1) dorsal view; O–P, MGUH30857 pygidium, GGU 274689 (CPL5); O, dorsal and P, oblique lateral views; Q–S, *Ceratocephala* sp. 2; Q–R, MGUH30858 cephalon; Q, dorsal and R, oblique lateral views; S, MGUH30859 hypostome, ventral view; both specimens from GGU 298506 (WL). All scale bars represent 1 mm.

proportionally and sculpturally, and from *C.* cf *C. bicuspis* by having lateral glabellar lobes less confluent with the fixigena so that they are better defined abaxially.

An associated fragmentary pygidium has two visible axial rings and a sculpture of centrally perforated tubercles which are more equal in size to one another than those of the cranidium. The pygidium bears three barbed marginal pygidial spines, separated by at least four marginal barbs and with at least four marginal barbs lateral to them. Chatterton and Perry (1983) noted that all *Ceratocephala* species from the Ordovician and early Silurian have three-spined pygidia, and the pygidium is best compared to the pygidia of *Ceratocephala* species from the Llandovery Whittaker Formation, Mackenzie Mountains, Canada, such as *C. avalanchensis* and *C. crawfordi* Chatterton and Perry, 1983 (p. 49, pl. 34, figs 18, 20–22, 26, 28, 30; text-fig. 42). The three pygidial border spines on these two species are separated by more than three marginal barbs, the condition seen in *C.* sp. 1.

Occurrence and distribution. The material is from Central Peary Land (CPL1–2, 4–5) with the exception of a cranidium from Kronprins Christian Land (KCL1).

Ceratocephala sp. 2
Figure 35Q–S

Remarks. Based on the limited material (a cephalon and a hypostome), similarities are evident with *C.* sp.

1, and species such as *C. laticapitata*, which share the medially expanded median glabellar lobe and well defined lateral glabellar lobes. The cephalon is differentiated from *C.* sp. 1 by the following: L1 and L2 are not distally fused and are narrower (tr.) and more oval in form; glabella and fixigenal sculpture comprises a smooth range in size from granules to larger centrally perforated tubercles, as opposed to the bimodal form of *C.* sp. 1, and these are randomly distributed as opposed to the regular pattern in *C.* sp. 1.

The librigena of *C.* sp. 2 exhibits the same sculpture as the glabella and fixigena, but the tubercles are more organized, occurring in rows aligned with the lateral border. The lateral border bears two rows of larger centrally perforated tubercles; it is narrower (tr.) and more convex posterior to the maximum width (tr.) of the librigena. The lateral border furrow is wide (tr.) and shallow, bordered adaxially by a row of centrally perforated tubercles which increase in size posteriorly.

The associated hypostome measures 2.3 mm maximum exsagittal length; 60% as long (exsag.) along maximum length, as wide (tr.) across shoulders, there positioned roughly at hypostomal mid-length; middle body 60% as long (sag.) as wide (tr.); anterior margin running transversely; middle furrow deep, only present laterally where directed posteromedially; lateral border furrow deepest posterior to middle furrow, and deeper here than posterior border furrow; posterior border abaxially inflated posteriorly.

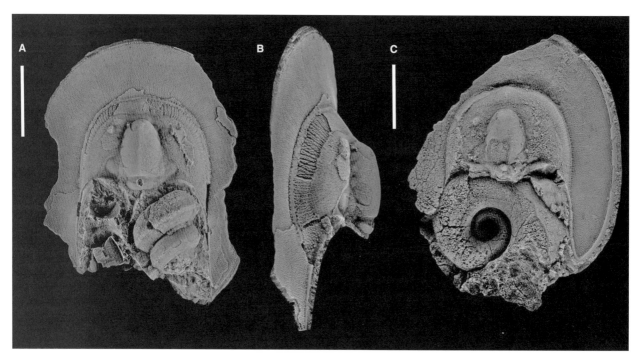

FIG. 36. *Scotoharpes loma* Lane, 1972. A–B, MGUH30860 cranidium; A, dorsal and B, lateral views. C, MGUH30861 cranidium, dorsal view. Both cranidia from GGU 275015 (KCL1). Both scale bars represent 10 mm.

Occurrence and distribution. The cephalon and hypostome are both from Wulff Land (WL).

Order Uncertain
Family HARPETIDAE Hawle and Corda, 1847

Remarks. Adrain (2011) considered the Family Harpetidae as Order Uncertain. The family had generally been included in the paraphyletic Order Ptychopariida previously, essentially for convenience.

Genus SCOTOHARPES Lamont, 1948

Type species. By monotypy; *Scotoharpes domina* Lamont, 1948, from the upper Llandovery Wether Law Linn Formation, North Esk Inlier, Scotland.

Other species. Scotoharpes aduncus Fortey, 1980; *S. cassinensis* (Whitfield, 1897); *S. consuetus* (Billings, 1866); *S. dalecarlicus* (Thorslund, 1930); *S. excavatus* (Linnarsson, 1875); *S. filiarum* Dean, 1979; *S. fragilis* (Raymond, 1925); *S. granti* (Billings, 1865); *S. judex* (Marr and Nicholson, 1888); *S. latior* (Poulsen, 1934); *S. lauriei* Jell and Stait, 1985; *S. lobulatus* (Chugaeva, 1975); *S. loma* (Lane, 1972); *S. meitanensis* Yin *in* Yin and Li, 1978; *S. molongloensis* Chatterton and Campbell, 1980; *S. pamiricus* (Balashova, 1966); *S. pansus* (Maximova, 1960); *S. raaschi* Norford, 1973; *S. rotundus* (Bohlin, 1955); *S. sanctacrucensis* (Kielan, 1959); *S. sinensis* (Grabau, 1925); *S. singularis* (Whittington, 1965); *S. sombrero* Owen, 1981; *S. spasskii* (Eichwald, 1840); *S. taimyricus* (Balashova, 1959); *S. tatouyangensis* (Chang and Fan, 1960); *S. telleri* (Weller, 1907); *S. tobulatus* (Chugaeva, 1975); *S. trinucleoides* (Etheridge and Mitchell, 1917); *S. vetustus* Zhou and Zhang, 1978; *S. vitilis* (Whittington, 1963); *S. volsellatus* Howells, 1982; *S. wegelini* (Angelin, 1854); *S. willsi* (Whittington, 1950); *S. youngi* (Reed, 1914).

Diagnosis. See Norford (1973, p. 11).

Scotoharpes loma Lane, 1972
Figure 36

v*. 1972 *Selenoharpes loma* Lane, pp. 353–355, pl. 62, figs 1–9.
. 1973 *Scotoharpes loma* Lane, Norford; pp. 18–20, pl. 3, figs 1–13.

Holotype. MMH 11357, cephalon, Lane, 1972 (p. 353, pl. 62, fig. 1a–b) from the Telychian Samuelsen Høj Formation of Kronprins Christian Land, eastern North Greenland.

Diagnosis. See Lane (1972, p. 353).

Remarks. The medial swelling on the cheek roll is particularly characteristic of the species. The present material differs in no significant respect from the type and figured material (see Lane (1972, p. 353) for full description). In our collections, this species is represented by some 61 cephala and one fragmentary thorax.

Occurrence and distribution. This is the most widely distributed of the taxa, being present in Wulff Land (WL), South Nares Land (SNL2), Western Peary Land (WPL1–2), Central Peary Land (CPL5), Kronprins Christian Land (KCL1) and Valdemar Glückstadt Land (VGL1). It is most abundant in South Nares Land.

Acknowledgements. This research was funded by a NERC PhD studentship to HEH (Grant no: NER/S/A/2006/14052) and undertaken at the University of Birmingham. Additionally, SYNTHESYS Funding enabled two trips to examine further collections of trilobites in Copenhagen. We thank Dr P. D. Lane for reading earlier and final versions of the manuscript and offering his advice and suggestions, and Dr R. M. Owens, Professor Derek J. Siveter and Dr D. J. Holloway for their comments on the proetaceans, Calymenidae and Scutelluidae, respectively. Professors J. M. Adrain and B. D. E. Chatterton, and Dr D. J. Holloway are thanked for their thorough reviews, and insightful comments for the improvement of the manuscript. Sten Lennart Jakobsen photographed the *O. magnifica* pygidia originally figured by Teichert, Dr C. M. Ø. Rasmussen kindly shared his knowledge of the Peel Collection, and John Clatworthy, Paul Hands and Aruna Mistry provided logistical support.

Editor. Andrew Smith

REFERENCES

ADRAIN, J. M. 1997. Proetid trilobites from the Silurian (Wenlock–Ludlow) of the Cape Phillips Formation, Canadian Arctic Archipelago. *Palaeontographica Italica*, **84**, 21–111.

—— 1998. Systematics of the Acanthoparyphinae (Trilobita), with species from the Silurian of Arctic Canada. *Journal of Paleontology*, **72**, 698–718.

—— 2003. Validity and composition of the Silurian trilobite genera *Borealarges* and *Dicranogmus*, with new species from the Canadian Arctic. *Canadian Journal of Earth Sciences*, **40**, 749–763.

—— 2011. Class Trilobita Walch. 1771. *In* ZHANG, Z.-Q. (ed.). *Animal biodiversity: an outline of higher-level classification and survey of taxonomic richness.* Zootaxa, **3148**, 104–109.

—— 2013. A synopsis of Ordovician trilobite distribution and diversity. 293–332. *In* HARPER, D. A. T. and SERVAIS, T. (eds). *Early Palaeozoic palaeobiogeography and palaeogeography.* Geological Society of London, Memoir, **38**, 490 pp.

—— and CHATTERTON, B. D. E. 1993. A new rorringtoniid trilobite from the Ludlow of Arctic Canada. *Canadian Journal of Earth Sciences*, **30**, 1634–1643.

—— —— 1995. Aulacopleurine trilobites from the Llandovery of Northwestern Canada. *Journal of Paleontology*, **69**, 326–340.

—— and EDGECOMBE, G. D. 1995. *Balizoma* and the new genera *Aegrotocatellus* and *Perirehaedulus*: encrinurid trilobites from the Douro Formation (Ludlow) of Arctic Canada. *Journal of Paleontology*, **69**, 736–752.

—— —— 1997. Silurian (Wenlock) calymenid trilobites from the Cape Phillips Formation, central Canadian Arctic. *Journal of Paleontology*, **71**, 657–682.

—— and FORTEY, R. A. 1997. Ordovician trilobites from the Tourmakeady Limestone, western Ireland. *Bulletin of the Natural History Museum, Geology Series*, **53**, 79–115.

—— and KLOC, G. J. 1997. Lower Devonian aulacopleuroidean trilobites from Oklahoma. *Journal of Paleontology*, **71**, 703–712.

—— and WESTROP, S. R. 2007. *Bearriverops*, a new Lower Ordovician trilobite genus from the Great Basin, western USA, and classification of the family Dimeropygidae. *Canadian Journal of Earth Sciences*, **44**, 337–366.

—— CHATTERTON, B. D. E. and BLODGETT, R. B. 1995. Silurian trilobites from Southwestern Alaska. *Journal of Paleontology*, **69**, 723–726.

ALBERTI, G. K. B. 1970. Trilobiten des jüngeren Siluriums sowie des Unter- und Mitteldevons. II. *Abhandlungen der Senckenbergischen Naturforschenden Gesellschaft*, **525**, 1–233, 20 pls.

—— 1981. Trilobiten des jüngeren Siluriums sowie des Unter- und Mittel- Devons. 3. Mit Beiträgen zur Devon-Biostratigraphie (insbesondere nach Nowakiidae) in N-Afrika. *Sardinien, Oberfranken und im Harz, Senckenbergiana Lethaea*, **62**, 1–75.

—— 2004. *Scharyia yolkiniana* nov. nom. instead of *Scharyia yolkini* Alberti 1983. *Senckenbergiana Lethaea*, **83**, 831–832.

ALDRIDGE, R. J. 1979. An upper Llandovery conodont fauna from Peary Land, eastern North Greenland. *Rapports Grønlands Geologiske Undersøgelse*, **91**, 7–23.

ANCYGIN, N. Ya. [ANTSYGIN, N. Ya.] 1973. [Trilobita]. 62–111. *In* VARGANOV, V. G., ANTSYGIN, N. Ya., NASEDKINA, V. A., MILITSINA, V. S. and SHURYGINA, M. V. (eds). *[Stratigraphy and faunas of the Ordovician of the Middle Urals]*. Nedra, Moscow, 228 pp., 30 pls. [In Russian]

—— 1977. [Trilobites of the Karakol-Mikhailovsk horizon of the Lower Ordovician of the southern Urals]. *Trudy Instituta Geologii i Geochimii, Akademia Nauk SSSR, Uralskii Nauchnyi Centr*, **126**, 65–95.

ANGELIN, N. P. 1854. *Palaeontologia Scandinavica. I. Crustacea formationis transitionis*, **Fascicule 2**, 21–92, pls 25–41, T. O. Weigel, Lund.

APOLLONOV, M. K. 1974. *[Ashgill trilobites from Kazakhstan]*. Akademiya Nauk Kazakh SSR, Alma Ata, 136 pp., 21 pls. [In Russian]

—— 1980. Klass Trilobita. 86–118. *In* APOLLONOV, M. K., BANDALETOV, S. M. and NIKITIN, I. F. (eds). *Granitsa Ordovika i Silura v Kazachstane*. Izdatelstvo Nauka Kazakhskoy SSR, Alma Ata, 300 pp. [In Russian]

ARMSTRONG, H. A. 1990. Conodonts from the Upper Ordovician – Lower Silurian carbonate platform of North Greenland. *Bulletin Grønlands Geologiske Undersøgelse*, **159**, 151 pp.

BALASHOVA, E. A. [BALAŠOVA, E. A.] 1959. Middle and Upper Ordovician and Lower Silurian trilobites of eastern Taimyr and their stratigraphical significance. *Sbornik Statei po Paleontologii I Biostratigrafii Institut Geologiya Arktiki*, **14**, 17–47.

—— 1966. [Trilobites from the Ordovician and Silurian deposits of Pamir]. *Trudy Upravleniâ Geologii Soveta Ministrov Tadžikskoj SSR. Paleontologiâ i Stratigraphiâ*, **2**, 191–262.

BARRANDE, J. 1846. *Notice préliminaire sur le Système silurien et les trilobites de Bohême*. Published by the author, Leipsic, vi + 97 pp.

—— 1852. *Système silurien du centre de la Bohême. 1ère partie: recherches paléontologiques. Vol. 1. Crustacés: Trilobites.* Published by the author, Prague and Paris, xxx + 935 pp., 51 pls.

—— 1872. *Système silurien du centre de la Bohême. 1ère partie: recherches paléontologiques. Supplément au vol. 1. Trilobites, crustacés divers et poissons.* Published by the author, Prague and Paris, xxx + 647 pp., 35 pls.

BARTON, D. C. 1916. A revision of the Cheirurinae, with notes on their evolution. *Washington University Studies, Scientific Series*, **3**, 101–152.

BEGG, J. L. 1939. Some new species of Proetidae and Otarionidae from the Ashgillian of Girvan. *Geological Magazine*, **76**, 372–382, pl. 6.

BEYRICH, E. 1845. *Ueber einige böhmischen Trilobiten*. G. Reimer, Berlin, 47 pp., 1 pl.

—— 1846. *Untersuchungen über Trilobiten. Zweites Stück. Als Forsetzung zu der Abhandlung 'Ueber einige böhmische Trilobiten'*. G. Reimer, Berlin, 37 pp., 4 pls.

BILLINGS, E. 1857. *Report for the year 1856 of E. Billings, Esq., Palaeontologist, addressed to Sir William. E. Logan, Provincial Geologist. Geological Survey of Canada, Report of Progress for the years 1853–54–55–56*, 247–345.

—— 1859. Description of some new species of trilobites from the Lower and Middle Silurian rocks of Canada. *Canadian Naturalist and Geologist*, **4**, 367–383.

—— 1860. Description of some new species of trilobites from the Lower and Middle Silurian rocks of Canada. *Canadian Naturalist and Geologist*, **5**, 49–73.

—— 1865. New species of fossils from the Quebec Group, in eastern Canada, with some others previously described, and some from other formations. *Geological Survey of Canada, Palaeozoic Fossils*, **1**, 301–388.

—— 1866. Catalogues of the Silurian fossils of the Island of Anticosti, with descriptions of some new genera and species. *Geological Survey of Canada, Separate Report*, **427**, 1–93.

—— 1869. Description of some new species of fossils with remarks on others already known, from the Silurian and Devonian rocks of Maine. *Proceedings of the Portland Society of Natural History*, **1**, 104–126.

BOHLIN, B. 1955. The Lower Ordovician limestones between the *Ceratopyge* Shale and the *Platyurus* Limestone of Böda Hamn with a description of the microlithology of the limestones by V. Jaanusson. *Bulletin of the Geological Institution of the University of Uppsala*, **35**, 111–173.

BRONGNIART, A. 1822. Les Trilobites. 1–65, pls 1–4. *In* BRONGNIART, A. and DESMAREST, A. G. (eds).

Histoire naturelle des crustacés fossiles. F.-G. Levrault, Paris, 154 pp., 11 pls.

BRUTON, D. L. 1966. A revision of the Swedish Ordovician Odontopleuridae (Trilobita). *Bulletin of the Geological Institution of the University of Uppsala,* **43,** 1–44.

BURMEISTER, H. 1843. *Die Organisation der Trilobiten, aus ihren lebenden Verwandten entwickelt; nebst einer systematischen Uebersicht aller zeither beschriebenen Arten.* Reimer, Berlin, 147 pp., 6 pls.

CAMPBELL, K. S. W. 1967. Trilobites of the Henryhouse Formation (Silurian) in Oklahoma. *Oklahoma Geological Survey, Bulletin,* **115,** 68 pp., 19 pls.

CAMPBELL, M. J. and CHATTERTON, B. D. E. 2006. Late Ordovician and Silurian Lichid trilobites from Northwestern Canada: eight new species from subfamilies Lichinae, Platylichinae, and Trochurinae. *Journal of Paleontology,* **80,** 514–528.

CARLUCCI, J. R., WESTROP, S. R., AMATI, L., ADRAIN, J. M. and SWISHER, R. E. 2012. A systematic revision of the Upper Ordovician trilobite genus *Bumastoides* (Illaenidae), with new species from Oklahoma, Virginia and Missouri. *Journal of Systematic Palaeontology,* **10,** 679–723.

CHANG, W.-T. 1974. *In* EDITORIAL BOARD OF NANJING INSTITUTE OF GEOLOGY AND PALAEONTOLOGY, ACADEMICA SINICA (ed.). *A handbook of stratigraphy and palaeontology of southwest China.* Academica Sinica, Academic and Science Press, Beijing, iii + 454 pp., 202 pls. [In Chinese]
—— and FAN, J.-S. 1960. [Ordovician and Silurian trilobites of the Chilian Mountains]. 83–148, pls 1–10. *In* YIN, T. (ed.). *Geological gazetteer of the Chilian mountains, Vol. 4.* Science Press, Beijing, 160 pp.

CHATTERTON, B. D. E. 1971. Taxonomy and ontogeny of Siluro-Devonian trilobites from Yass, New South Wales. *Palaeontographica,* A, **137,** 108 pp.
—— and CAMPBELL, K. S. W. 1980. Silurian trilobites from near Canberra and some related forms from the Yass Basin. *Palaeontographica,* A, **167,** 77–119.
—— and LUDVIGSEN, R. 2004. Silurian trilobites of Anticosti Island, Québec, Canada. *Palaeontographica Canadiana,* **22,** 264 pp., 85 pls.
—— and PERRY, D. G. 1983. Silicified Silurian odontopleurid trilobites from the Mackenzie Mountains. *Palaeontographica Canadiana,* **1,** 1–127, 36 pls.
—— —— 1984. Silurian cheirurid trilobites from the Mackenzie Mountains, northwestern Canada. *Palaeontographica,* A, **184,** 1–78.
—— EDGECOMBE, G. D., WAISFELD, B. G. and VACCARI, N. E. 1998. Ontogeny and systematics of Toernquistiidae (Trilobita, Proetida) from the Ordovician of the Argentine Precordillera. *Journal of Paleontology,* **72,** 273–303.

CHERNYSHEV, A. 1893. The fauna of the Lower Devonian on the western slope of the Urals. *Mémoires du Comité Géologique,* **4,** 1–221. [In Russian and German]

CHLUPÁČ, I. 1971. Some trilobites from the Silurian/Devonian boundary beds of Czechoslovakia. *Palaeontology,* **14,** 159–177, pls 19-24.
—— 1987. Ecostratigraphy of Silurian trilobite associations of the Barrandian area, Czechoslovakia. *Newsletters on Stratigraphy,* **17,** 169–186.

CHRISTIE, R. L. and PEEL, J. S. 1977. Cambrian–Silurian stratigraphy of Børglum Elv, Peary Land, eastern North Greenland. *Rapport Grønlands Geologiske Undersøgelse,* **82,** 1–48.

CHUGAEVA, M. N. 1975. Late Ordovician trilobites of the northeast of the USSR. *Trudy Geologicheskogo Instituta, Akademiya Nauk SSSR,* **272,** 1–75, pls 1–4.
—— 1983. [Systematic description of trilobites]. 73–97. *In* KOREN', T. N., ORADOVSKAYA, M. M., PYLMA, L. J., SOBOLEVSKAYA, R. F. and CHUGAEVA, M. N. (eds). *Granica Ordovika i Silura na Severo-Vostoke SSSR [The Ordovician and Silurian Boundary in the northeast of the USSR].* Trudy Akademiya Nauk SSSR, Minsterstvo Geologii SSSR, Mezhvedomstvennyj Stratigrafcheskij Komitet SSSR, **11,** 204 pp. [In Russian]

CLARKSON, E. N. K. and HOWELLS, Y. 1981. Upper Llandovery trilobites from the Pentland Hills, near Edinburgh, Scotland. *Palaeontology,* **24,** 507–536.

COCKS, L. R. M. and TORSVIK, T. H. 2002. Earth geography from 500 to 400 million years ago: a faunal and palaeomagnetic review. *Journal of the Geological Society of London,* **159,** 631–644.

COOPER, G. A. 1930. Upper Ordovician and Lower Devonian stratigraphy and paleontology of Percé, Quebec. II. New species from the Upper Ordovician of Percé. *American Journal of Science,* **20** (265–288), 365–392.
—— and KINDLE, C. H. 1936. New brachiopods and trilobites from the Upper Ordovician of Percé, Quebec. *Journal of Paleontology,* **10,** 348–372, pls 51–53.

COPPER, P. and BRUNTON, F. 1991. A global review of Silurian reefs. *Special Papers in Palaeontology,* **44,** 225–259.

CURTIS, M. L. K. 1958. The upper Llandovery trilobites of the Tortworth inlier, Gloucestershire. *Palaeontology,* **1,** 139–146, pl. 29.

CURTIS, N. J. and LANE, P. D. 1997. The Llandovery trilobites of England and Wales. Part 1. *Palaeontographical Society Monograph,* **151,** 1–50.

DALMAN, J. W. 1827. Om palaeaderna, eller de så kallade trilobiterna. *Kungliga Svenska Vetenskaps-Akademiens Handlingar* (for 1826), 113–152, 226–294, 6 pls.

DAWES, P. R. 1971. The North Greenland fold belt and environs. *Meddelelser Dansk Geologiske Forening,* **20,** 197–239.
—— 1976. Precambrian to Tertiary of northern Greenland. 248–303. *In* ESCHER, A. and WATT, W. S. (eds). *Geology of Greenland.* Geological Survey of Greenland, Copenhagen, 603 pp.

DEAN, W. T. 1979. Trilobites from the Long Point Group (Ordovician), Port au Port Peninsula, south-western Newfoundland. *Geological Survey of Canada, Bulletin,* **290,** 1–53.

DESMAREST, A. G. 1817. *Nouveau Dictionnaire d'Histoire Naturelle.* Second edition, Vol. 5. Deterville, Paris, 49–50.

EDGECOMBE, G. D. and ADRAIN, J. M. 1995. Silurian calymenid trilobites from the United States. *Palaeontographica,* A, **235,** 1–19.
—— and CHATTERTON, B. D. E. 1992. Early Silurian (Llandovery) encrinurine trilobites from the Mackenzie Mountains, Canada. *Journal of Paleontology,* **66,** 52–74.
—— —— 1993. Silurian (Wenlock–Ludlow) encrinurine trilobites from the Mackenzie Mountains, Canada, and related species. *Palaeontographica Abteilung,* A, **229,** 75–112.

—— and RAMSKÖLD, L. 1996. The '*Encrinurus*' *variolaris* plexus (Trilobita, Silurian): relationships of Llandovery species. *Geobios*, **29**, 209–233.

—— and WRIGHT, A. J. 2004. Silicified Early Devonian trilobites from Brogans Creek, New South Wales. *Proceedings of the Linnean Society of New South Wales*, **125**, 177–188.

—— BANKS, M. R. and BANKS, D. M. 2006. *Bumastoides* (Trilobita: Illaenidae) from the Upper Ordovician of Tasmania. *Memoirs of the Association of Australasian Palaeontologists*, **32**, 375–381.

EICHWALD, K. E. 1840. *Über das silurische Schichten-system von Esthland.* St. Petersburg Academy, 210 pp.

ERBEN, H. K. 1951. Beitrag zur Gliederung der Gattung *Proetus* Stein., 1831 (Trilobitae). *Neues Jahrbuch für Geologie und Paläontologie, Abhandlungen*, **94**, 5–48.

ETHERIDGE, R. Jr 1896. Description of a small collection of Tasmanian Silurian fossils presented to the Australian Museum by Mr. A. Montgomery, M.A., Government Geologist, Tasmania. *Report of the Secretary for Mines for 1895–6, Tasmania*, 61–67.

—— and MITCHELL, J. 1896. The Silurian trilobites of New South Wales, with references to those of other parts of Australia. Part IV. The Odontopleuridae. *Proceedings of the Linnean Society of New South Wales*, **11**, 694–721, pls 50–55.

—— —— 1917. The Silurian trilobites of New South Wales, with references to those of other parts of Australia. Part 6. The Calymenidae, Cheiruridae, Harpeidae, Bronteidae, etc., with an appendix. *Proceedings of the Linnean Society of New South Wales*, **42**, 480–510.

FOERSTE, A. F. 1887. The Clinton Group of Ohio. Part II. *Bulletin of the Scientific Laboratories of Denison University*, **2**, 89–110, pl. 8.

FORTEY, R. A. 1975. Early Ordovician trilobite communities. *Fossils and Strata*, **4**, 331–352.

—— 1980. The Ordovician trilobites of Spitzbergen III. Remaining trilobites of the Valhallfonna Formation. *Norsk Polarinstitutt Skrifter*, **171**, 1–163.

—— and COCKS, L. R. M. 2003. Palaeontological evidence bearing on global Ordovician–Silurian continental reconstructions. *Earth Science Reviews*, **61**, 245–307.

—— and OWENS, R. M. 1975. Proetida: a new order of trilobites. *Fossils and Strata*, **4**, 227–239.

FRITZ, M. A. 1964. ?*Scutellum regale* sp. nov. Fritz from the Silurian of the Hudson Bay area. *Proceedings of the Geological Association of Canada*, **15**, 91–97.

GASS, K. C. and MIKULIC, D. G. 1982. Observations on the Attawapiskat Formation (Silurian) trilobites of Ontario, with description of a new encrinurine. *Canadian Journal of Earth Sciences*, **19**, 589–596.

GOLDFUSS, A. 1843. Systematische Übersicht der Trilobiten und Beschreibung einiger neue Arten derselben. *Neues Jahrbuch füer Mineralogie, Geognosie, Geologie und Petrefaktenkunde* (for 1843) Berlin, 537–567, 3 pls.

GRABAU, A. W. 1925. Summary of the faunas from the Sintan shale. *Bulletin of the Geological Survey of China*, **7**, 77–85.

HALL, J. 1861. *Natural History of New York. Palaeontology* (for 1859). Albany, **3**, 532, 120 pls.

—— 1864. Account of some new or little-known species of fossils from rocks of the age of the Niagara Group. *Report of the New York State Museum of Natural History*, **18**, 1–48, 1 pl.

—— 1867. *Report of the New York State Museum of Natural History*, **20**, 305–401 (Republication, with additions, of Hall 1864b).

HAMILTON, K. G. A. 1990. Insects from the Santana Formation, Lower Cretaceous, of Brazil, Homoptera. *Bulletin of the American Museum of Natural History*, **195**, 82–122.

HARPER, D. A. T., MACNIOCAILL, C. and WILLIAMS, S. H. 1996. The palaeogeography of Early Ordovician Iapetus terranes: an integration of faunal and palaeomagnetic constraints. *Palaeogeography, Palaeoclimatology, Palaeoecology*, **121**, 297–312.

HARRINGTON, H. J. and LEANZA, A. F. 1957. Ordovician trilobites of Argentina. *Special Publications of the Department of Geology, University of Kansas*, **1**, 1–276, 140 figs.

HAWLE, I. and CORDA, A. J. C. 1847. *Prodrom einer Monographie der böhmischen Trilobiten.* J. G. Calvesche, Prague, 176 pp., 7 pls.

HEDSTRÖM, H. 1923. Contributions to the fossil fauna of Gotland. I. *Sveriges Geologiska Undersökning, C*, **316**, 1–25, 5 pls.

HENRIKSEN, N., HIGGINS, A. K., KALSBEEK, F. and PULVERTAFT, T. C. R. 2009. *Greenland from Archaean to Quaternary. Descriptive text to the 1995 Geological map of Greenland 1:2 500 000*, Second edition. Geological Survey of Denmark and Greenland, Copenhagen, 126 pp.

HIGGINS, A. K., INESON, J. R., PEEL, J. S., SURLYK, F. and SØNDERHOLM, M. 1991. The Franklinian Basin in North Greenland. *Bulletin Grønlands Geologiske Undersøgelse*, **160**, 71–139.

HOLLOWAY, D. J. 1980. Middle Silurian trilobites from Arkansas and Oklahoma, U.S.A. *Palaeontographica Abteilung, A*, **170**, 1–85.

—— 2007. The trilobite *Protostygina* and the composition of the Styginidae, with two new genera. *Paläontologische Zeitschrift*, **81**, 1–16.

—— and LANE, P. D. 1998. Effaced styginid trilobites from the Silurian of New South Wales. *Palaeontology*, **41**, 853–896.

—— —— 1999. A replacement name for the trilobite *Lalax* Holloway and Lane *non* Hamilton. *Palaeontology*, **42**, 375.

—— —— 2012. Scutelluid trilobites from the Silurian of New South Wales. *Palaeontology*, **55**, 413–490.

HOLM, G. 1882. De svenska arterna af trilobitslägtet *Illaenus* (Dalman). *Bihang till Kungliga Svenska Vetenskaps-Akademiens Handlingar*, **7**, i–xiv + 1–148.

—— 1886. Revision der ostbaltischen silurischen Trilobiten. Abtheilung III. Illaeniden. *Mémoirs de l'Academie Impériale des Sciences de St Pétersbourg*, **33**, 1–173, 12 pls.

HÖRBINGER, F. 2004. Trilobites from the biodetritic facies of the lower part of Lochkov Formation (Lochkovian, Lower Devonian) from "Požáry-Vokounka" quarry (Prague-Řeporyje, Prague Basin, Czech Republic). *Palaeontologia Bohemiae*, **9**, 19–31 (privately published).

HOWELLS, Y. 1982. Scottish Silurian trilobites. *Palaeontographical Society Monograph*, **135**, 1–76, 15 pls.

HUGHES, H. E. and THOMAS, A. T. 2011. Trilobite associations, taphonomy, lithofacies and environments of the Silurian reefs of North Greenland. *Palaeogeography Palaeoclimatology Palaeoecology*, **302**, 142–155.

HUPÉ, P. 1953. Classe de trilobites. 44–246. *In* PIVETEAU, J. (ed.). *Traité de paleontologié, Vol. 3.* Masson and Cie, Paris, 1064 pp.

HURST, J. M. 1980. Palaeogeographic and stratigraphic differentiation of Silurian carbonate buildups and biostromes of North Greenland. *The American Association of Petroleum Geologists Bulletin*, **64**, 527–548.

—— 1984. Upper Ordovician and Silurian carbonate shelf stratigraphy, facies and evolution, eastern North Greenland. *Bulletin Grønlands Geologiske Undersøgelse*, **148**, 1–70.

INGHAM, J. K. and TRIPP, R. P. 1991. The trilobite fauna of the Middle Ordovician Doularg Formation of the Girvan District, Scotland, and its paleoenvironmental significance. *Transactions of the Royal Society of Edinburgh: Earth Sciences*, **82**, 27–54.

IVANOVA, O., OWENS, R. M., KIM, I. and POPOV, L. E. 2009. Late Silurian trilobites from the Nuratau and Turkestan ranges, Uzbekistan and Tajikistan. *Geobios*, **42**, 715–737.

JAANUSSON, V. 1954. Zur Morphologie und Taxonomie de Illaeniden. *Arkiv för Mineralogi och Geologi*, **1**, 545–583, 3 pls.

—— 1957. Zur Morphologie und Taxonomie der Illaeniden. *Arkiv för mineralogi och geologi*, **1**, 545–583, 3 pls.

JELL, P. A. and ADRAIN, J. M. 2003. Available generic names for trilobites. *Memoirs of the Queensland Museum*, **48**, 331–553.

—— and STAIT, B. 1985. Tremadoc trilobites from the Florentine Valley Formation, Tim Shea area, Tasmania. *Memoirs of the Museum of Victoria*, **46**, 1–34.

KEGEL, W. 1927. Über obersilurische Trilobiten aus dem Harz und dem Rheinischen Schiefergebirge. *Jahrbuch der Preussischen Geologischen Landesanstalt und Bergakademie zu Berlin*, **48**, 616–647, pls 31–32.

KIÆR, J. 1908. Das Obersilur im Kristianagebiete. Eine stratigraphisch-faunistische Untersuchung. *Skrifer Utgitt af Videnskabs Selskabet i Christiania, Matematisk-Naturvidenskapelig Klasse* (for 1906), **2**, i–xvi + 1–596.

KIELAN, Z. 1959. Upper Ordovician trilobites from Poland and some related forms from Bohemia and Scandinavia. *Palaeontologia Polonica*, **11**, 1–198, pls 1–36.

KJERULF, R. 1865. Veiviser ved geologisker Excursioner I Christiania Omegn. *Universitets-Program Seminar*, **2**, 43 pp.

KOBAYASHI, T. 1935. The Cambro-Ordovican Formations and Faunas of south Chosen. Palaeontology, Part 3, Cambrian Faunas of south Chosen with a special study on the Cambrian trilobite genera and families. *Journal of the Faculty of Science*, **4**, 49–344.

—— and HAMADA, T. 1977. Devonian trilobites of Japan in comparison with Asian pacific and other faunas. *Palaeontological Society of Japan, Special Papers*, **20**, 1–202.

—— —— 1984. Advance report on a new trilobite collection of the Silurian Yokokura-yama fauna, Shikoku island, Japan. *Research Report of the Kôchi University, Natural Sciences*, **32**, 253–258.

—— —— 1986. The second addition to the Silurian trilobite fauna of Yokokura-yama, Shikoku, Japan. *Transactions and Proceedings of the Palaeontological Society of Japan, New Series*, **143**, 447–462, pls 90–92.

—— —— 1987. The third addition to the Silurian trilobite fauna of Yokokura-yama, Shikoku, Japan. *Transactions and Proceedings of the Palaeontological Society of Japan, New Series*, **147**, 109–116.

LAMONT, A. 1948. Scottish dragons. *Quarry Managers' Journal*, **31**, 531–535.

LANE, P. D. 1971. British Cheiruridae (Trilobita). *Palaeontographical Society Monograph*, **125**, 1–95, 16 pls.

—— 1972. New trilobites from the Silurian of north-east Greenland, with a note on trilobite faunas in pure limestones. *Palaeontology*, **15**, 336–364.

—— 1979. Llandovery trilobites from Washington Land, North Greenland. *Bulletin Grønlands Geologiske Undersøgelse*, **131**, 1–37, 6 pls.

—— 1984. Silurian trilobites from Hall Land and Nyeboe Land, western North Greenland. *Rapports Grønlands Geologiske Undersøgelse*, **121**, 51–73, 4 pls.

—— 1988. Silurian trilobites from Peary Land, central North Greenland. *Rapports Grønlands Geologiske Undersøgelse*, **137**, 93–117.

—— and OWENS, R. M. 1982. Silurian trilobites from Kap Schuchert, Washington Land, western North Greenland. *Rapports Grønlands Geologiske Undersøgelse*, **107**, 41–69, pls 1–5.

—— and SIVETER, D. J. 1991. A Silurian trilobite fauna dominated by *Calymene* from Kap Tyson, Hall Land, western North Greenland. *Rapports Grønlands Geologiske Undersøgelse*, **150**, 5–14, 2 pls.

—— and THOMAS, A. T. 1978. Silurian trilobites from NE Queensland and the classification of effaced trilobites. *Geological Magazine*, **115**, 351–358.

—— —— 1979. Silurian carbonate mounds in Peary Land, north Greenland. *Rapports Grønlands Geologiske Undersøgelse*, **88**, 51–54.

—— —— 1983. A review of the trilobite suborder Scutelluinae. *Special Papers in Palaeontology*, **30**, 141–160.

LINDSTRÖM, G. 1885. Förteckning på Gotlands Siluriska crustacéer. *Öfversigt af Kungliga Vetenskaps-Akademiens Förhandlingar*, **42**, 37–100, pls 12–16.

LINNARSSON, J. G. O. 1875. Öfversigt af Nerikes öfvergångsbildningar. *Öfversigt af Kungliga Vetenskaps-Akademiens Förhandlingar*, **5**, 3–47, pls 4–5.

LOWENSTAM, H. A. 1950. Niagaran reef in the Great Lakes area. *Journal of Geology*, **58**, 430–487.

—— 1957. Niagaran reef in the Great Lakes area. *Memoir of the Geological Society of America*, **67**, 215–248.

LU, Y. H. 1975. Ordovician trilobite faunas of central and southwestern China. *Palaeontologica Sinica (New Series B)*, **10**, 1–484.

LUDVIGSEN, R. 1979. *Fossils of Ontario, Part 1: the Trilobites.* Royal Ontario Museum, Toronto, Life Sciences Miscellaneous Publications, 1–96.

—— and CHATTERTON, B. D. E. 1980. The ontogeny of *Failleana* and the origin of the Bumastinae (Trilobita). *Geological Magazine*, **117**, 471–478.

—— and TRIPP, R. P. 1990. Silurian trilobites from the northern Yukon Territory. *Royal Ontario Museum, Life Sciences Contributions*, **153**, 1–59.

LÜTKE, F. 1990. Contributions to a phylogenetical classification of the subfamily Proetinae Salter, 1864 (Trilobita). *Senckenbergiana lethaea*, **71**, 1–83.

MABILLARD, J. E. 1980. Silurian carbonate mounds of south-east Peary Land, eastern North Greenland. *Rapports Grønlands Geologiske Undersøgelse*, **99**, 57–60.

MACNIOCAILL, C., VAN DER PLUIJM, B. A. and VAN DER VOO, R. 1997. Ordovician paleogeography and the evolution of the Iapetus Ocean. *Geology*, **25**, 159–162.

MAKSIMOVA, Z. A. 1955. [Trilobites of the Middle and Upper Devonian of the Urals and northern Mugodzhar]. *Trudy Vsesoyuznogo Nauchno-Issledovatel' Skogo Geologicheskogo Instituta (VSEGEI), New Series*, **3**, 1–263, pls 1–18. [in Russian]

—— 1962. Trilobity Ordovika I Silua Sibirskoj platformy. [Ordovician and Silurian trilobites of the Siberian Platform]. *Trudy Vsesoyuznogo Nauchno-Issledovatel' Skogo Geologischeskogo Instituta (VSEGEI)*, **76**, 215 pp. [In Russian]

—— 1975. [Trilobites]. 119–133. *In* MENNER, V. V. (ed.). *Kharakteristika fauny pogranichnikh sloev Silura i Devona tsentral'nogo Kazakhstana. [Characteristic fauna of the boundary beds between the Silurian and Devonian of central Kazakhstan]* Materialy po geologii tsentral'nogo Kazakhstana, **12**, Nedra, Moscow, 248 pp. [in Russian]

MÄNNIL, R. 1958. Trilobity semeystv Cheiruridae i Encrinuridae iz Estonii. *Eesti NSV Teaduste Akadeemia Geoloogia Instituudi Uurimused*, **3**, 165–212.

—— 1982a. Wenlock and late Silurian trilobite associations of the east Baltic area and their stratigraphical value. 63–70. *In* KALJO, D. and KLAAMANN, E. (eds). *Ecostratigraphy of the East Baltic Silurian*. Valgus, Tallinn, 109 pp.

—— 1982b. Trilobite communities (Wenlock, East Baltic). 51–62. *In* KALJO, D. L. and KLAAMANN, E. (eds). *Communities and biozones in the Baltic Silurian*. Valgus, Tallinn, 109 pp. [In Russian]

MARR, J. E. and NICHOLSON, H. A. 1888. The Stockdale Shales. *Quaterly Journal of the Geological Society of London*, **44**, 654–732, pl. 16.

MAYR, U. 1976. Middle Silurian reefs in southern Peary Land, North Greenland. *Bulletin of Canadian Petroleum Geology*, **24**, 440–449.

MCCOY, F. 1846. *A synopsis of the Silurian fossils of Ireland*. Dublin University Press, 72 pp.

MIKULIC, D. G. 1981. Trilobites in Palaeozoic carbonate buildups. *Lethaia*, **14**, 45–56.

—— 1999. Silurian trilobite associations in North America. 793–798. *In* BOUCOT, A. J. and LAWSON, J. D. (eds). *Paleocommunities: a case study from the Silurian and Lower Devonian*. Cambridge University Press, Cambridge, 895 pp.

MOORE, R. C. 1959. *Treatise on invertebrate paleontology. Part O. Arthropoda 1*. Geological Society of America, Boulder, Colorado and University of Kansas Press, Lawrence, Kansas, xix + 560 pp.

MORRIS, S. F. 1988. A review of British trilobites, including a synoptic revision of Salter's monograph. *Palaeontographical Society Monograph*, **140**, 1–316.

MURCHISON, R. I. 1839. *The Silurian System, founded on geological researches in the counties of Salop, Hereford, Radnor, Montgomery, Caermarthen, Brecon, Pembroke, Monmouth, Gloucester, Worcester and Stafford; with descriptions of the coalfields and overlying formations*. John Murray, London, xxxii + 768 pp., 37 pls.

NAN, R.-S. 1976. Trilobites. 333–351, pls 195–201. *In Palaeontological Atlas of north China, Part 1, Inner Mongolia*. Geological Publishing House, Beijing, 502 pp.

NORFORD, B. S. 1973. Lower Silurian species of the trilobite *Scotoharpes* from Canada and Northwestern Greenland. *Geological Survey of Canada, Bulletin*, **479**, 13–47.

—— 1981. The trilobite fauna of the Attawapiskat Formation, northern Ontario and northern Manitoba. *Geological Survey of Canada, Bulletin*, **327**, 1–37.

—— 1994. Biostratigraphy and trilobite fauna of the Lower Silurian Tegart Formation, southeastern British Columbia. *Geological Survey of Canada, Bulletin*, **479**, 13–47.

OGG, J. G., OGG, G. and GRADSTEIN, F. M. 2008. *The concise geologic time scale*. Cambridge University Press, 150 pp.

ÖPIK, A. A. 1937. Trilobiten aus Estland. *Acta et Commentationes Universitatis Tartuensis, A*, **32**, 1–163.

OSMÓLSKA, H. 1957. Trilobites from the Couvinian of Wydryszów (Holy Cross Mountains, Poland). *Acta Palaeontologica Polonica*, **2**, 53–77, 3 pls.

—— 1970. Revision of non-cyrtosymbolinid trilobites from the Tournaisian–Namurian of Eurasia. *Palaeontologia Polonica*, **23**, 1–165, 22 pls.

OVER, D. J. and CHATTERTON, B. D. E. 1987. Silurian conodonts from the southern Mackenzie Mountains, Northwest Territories, Canada. *Geologica et Palaeontologica*, **21**, 1–49.

OWEN, A. W. 1981. The Ashgill trilobites of the Oslo Region, Norway. *Palaeontographica, A*, **175**, 1–88, 17 pls.

OWENS, R. M. 1973a. British Ordovician and Silurian Proetidae (Trilobita). *Palaeontographical Society Monograph*, **127**, 1–98, 15 pls.

—— 1973b. Ordovician Proetidae (Trilobita) from Scandinavia. *Norsk Geologisk Tidsskrift*, **53**, 117–181.

—— 1974. The affinities of the trilobite genus *Scharyia*, with a description of two new species. *Palaeontology*, **17**, 685–697.

—— 2006. The proetid trilobite *Hedstroemia* and related Ordovician to Carboniferous taxa. 119–143. *In* BASSETT, M. G. and DEISLER, V. K. (eds). *Studies in Palaeozoic palaeontology*. National Museum of Wales Geological Series No. 25, Cardiff, 294 pp.

—— and FORTEY, R. A. 2009. Silicified Upper Ordovician trilobites from Pai-Khoi, Arctic Russia. *Palaeontology*, **52**, 1209–1220.

—— and HAMMANN, W. 1990. Proetide trilobites from the Cystoid Limestone (Ashgill) of NW Spain, and the suprageneric classification of related forms. *Paläontologische Zeitschrift*, **64**, 221–244.

—— IVANOVA, O., KIM, I., POPOV, L. E. and FEIST, R. 2010. Lower and Middle Devonian trilobites from southern

Uzbekistan. *Memoirs of the Association of Australasian Palaeontologists*, **39**, 211–244.

PEEL, J. S. and SØNDERHOLM, M. (eds). 1991. Sedimentary basins of North Greenland. *Bulletin Grønlands Geologiske Undersøgelse*, **160**, 164 pp.

PENG, S.-C. 1990. Tremadoc stratigraphy and trilobite faunas of northwestern Huna. *Beringeria*, **2**, 1–171.

PERRY, D. G. and CHATTERTON, B. D. E. 1977. Silurian (Wenlockian) trilobites from Baillie-Hamilton Island, Canadian Arctic Archipelago. *Canadian Journal of Earth Sciences*, **14**, 285–317.

—— —— 1979. Wenlock trilobites and brachiopods from the Mackenzie Mountains, north-western Canada. *Palaeontology*, **22**, 569–607.

PHILLIPS, J. and SALTER, J. W. 1848. Palaeontological appendix to Professor John Phillips' memoir on the Malvern Hills compared with the Palaeozoic Districts of Abberley etc. *Memoir of the Geological Survey of Great Britain*, **2**, 331–386, 20 pls.

PHLEGER, F. B. 1936. Lichadian trilobites. *Journal of Paleontology*, **10**, 593–615.

POULSEN, C. 1934. The Silurian faunas of North Greenland, I. The fauna of the Cape Schuchert Formation. *Meddelelser om Grønland*, **72**, 1–47.

PRANTL, F. and PŘIBYL, A. 1949. A study of the superfamily Odontopleuracea nov. superfam. (trilobites). *Rozpravy Státního Geologického Ústavu, ČSR*, **12**, 1–221.

PŘIBYL, A. 1946a. Příspěvek k poznání českých proetidů. *Rozpravy České Akademie věd a Umění, Třída 2 Prague*, **55**, 1–37.

—— 1946b. O několika nových trilobitových rodech z českého siluru a devonu. *Příroda*, **38**, 89–95, 7 figs.

—— 1966. Proetidní trilobiti z nových sběrů v českém siluru a devonu. [Proetiden aus neueren Aufsammlungen im böhmischen Silur und Devon (Trilobitae) II.]. *Časopis národního Muzea*, **135**, 49–54, pl. 4. [In Czech and German]

—— 1967. Die Gattung *Scharyia* Přibyl, 1946 (Trilobita) und ihre Vertreter aus dem böhmischen Silur und Devon. *Spisanie na Bŭlgarskoto Geologichesko Druzhestvo*, **28**, 285–301.

—— 1970. O několika českých a asijských zástupcích proetidních trilobitů. [Über einige bohemische und asiatische Vertreter von Proetiden (Trilobita)]. *Časopis pro Mineralogii a Geologii*, **15**, 101–111, 1 pl. [In German with Czech summary]

—— and VANĚK, J. 1971. Studie über die Familie Scutelluidae Richter et Richter (Trilobita) und ihre phyllogenetische Entwicklung. *Acta Universitatis Carolinae, Geologica*, **1971** (4), 361–394.

—— —— 1980. Studie zu einigen neuen Trilobiten der Proetidae-Familie. *Acta Universitatis Carolinae, Geologica*, **1978**, 163–182.

—— —— and HÖRBINGER F. 1985. New taxa of Proetacea (Trilobita) from the Silurian and Devonian of Bohemia. *Časopis pro Mineralogii a Geologii*, **30**, 237–251.

PROUTY, W. F. 1923. Trilobita. 704–715. *In* SWARTZ, C. K., PROUTY, W. F., ULRICH, E. O. and BASSLER, R. S. (eds). *Systematic paleontology of Silurian deposits*. Maryland Geological Survey, Silurian, 794 pp.

QIU, H.-A., LU, Y.-H., ZHU, Z.-L., BI, D.-C., LIN, T.-R., ZHOU, Z.-Y., ZHANG, Q.-H., QIAN, Y.-Y., JU, T.-Y., HAN, N.-R. and WEI, X.-Z. 1983. [Trilobita]. 28–254. *In Paleontological Atlas of East China. Part 1: Early Paleozoic.* Nanjing Institute of Geology and Mineral Resources, Geological Publishing House, Beijing, 857 pp.

RAMSKÖLD, L. 1983. Silurian cheirurid triloites from Gotland. *Palaeontology*, **26**, 175–210.

—— 1986. Silurian encrinurid trilobites from Gotland and Dalarna, Sweden. *Palaeontology*, **29**, 527–575.

—— 1988. Heterochrony in Silurian phacopid trilobites as suggested by the ontogeny of *Acernaspis*. *Lethaia*, **21**, 307–318.

—— 1991. Pattern and process in the evolution of the Odontopleuridae (Trilobita). The Selenopeltinae and Ceratocephalinae. *Transactions of the Royal Society of Edinburgh: Earth Sciences*, **82**, 143–181.

—— and WERDELIN, L. 1991. The phylogeny and evolution of some phacopid trilobites. *Cladistics*, **7**, 29–74.

RAYMOND, P. E. 1916. New and old Silurian trilobites from south-eastern Wisconsin, with notes on the genera of the Illaenidae. *Bulletin of the Museum of Comparative Zoology, Harvard College*, **60**, 1–41.

—— 1925. Some trilobites of the lower Middle Ordovician of eastern North America. *Bulletin of the Museum of Comparative Zoology*, **67**, 1–180.

REED, F. R. C. 1896. The fauna of the Keisley Limestone, Part I. *Quarterly Journal of the Geological Society of London*, **52**, 407–437, pls 20–21.

—— 1904. The Lower Palaeozoic trilobites of the Girvan district, Ayrshire. Part 2. *Palaeontographical Society Monograph*, **58** (276), 49–96, pls 7–13.

—— 1914. The Lower Palaeozoic trilobites of Girvan. Supplement. *Palaeontographical Society Monograph*, **66** (329), 56 pp., 8 pl.

—— 1933. Notes on the species *Illaenus bowmanni* Salter. *Geological Magazine*, **70**, 121–135.

—— 1935. The Lower Palaeozoic trilobites of Girvan. Supplement No 3. *Palaeontographical Society Monograph*, **88** (400), 1–64, 4 pls.

—— 1941. A new genus of trilobites and other fossils from Girvan. *Geological Magazine*, **78**, 268–278, pl. 5.

RICHTER, R. and RICHTER, E. 1917. Über die Einteilung der Familie Acidaspidae und über einige ihrer devonischer Vertreter. *Zentralblatt für Mineralogie, Geologie und Paläontologie* (for 1917), 462–472, 10 figs.

—— —— 1925. Unterlagen zum Fossilium Catalogus, Trilobita. II. *Senckenbergiana lethaea*, **7**, 126.

—— —— 1955. Scutelluidae n. n. (Tril.) durch 'kleine Änderung' eines Familien-Namens wegen Homonymie. *Senckenbergiana lethaea*, **36**, 291–293.

ROSENSTEIN, E. 1941. Die *Encrinurus* – Arten des Estländischen Silurs. *Annales Societatis Rebus Naturae Investigandis in Universitate Tartuensis Consitutae*, **47**, 49–77.

SALTER, J. W. 1853. Figures and descriptions illustrative of British organic remains. *Memoir of the Geological Survey of the United Kingdom*, **December 7**, 121 pp., pl. 2.

—— 1864. A monograph of the British trilobites from the Cambrian, Silurian and Devonian formations. *Palaeontographical Society Monograph*, **16**, 1–80, 6 pls.

—— 1867. A monograph of the British trilobites from the Cambrian, Silurian and Devonian formations. *Palaeontographical Society Monograph*, **20**, 177–214, 6 pls.

SCHMIDT, F. 1894. Revision der ostbaltischen silurischen Trilobiten. Abtheilung IV. Calymmeniden, Proetiden, Bronteiden, Harpediden, Trinucleiden, Remopleuriden und Agnostiden. *Mémoires de l'Académie des Sciences de St.-Pétersbourg*, VIIe Série, Tome XLII, no. 5. iii + 93 pp., 6 pls.

SCHRANK, E. 1972. Proetacea, Encrinuridae und Phacopina (Trilobita) aus silurischen Geschieben. *Geologie*, **76**, 1–117, 21 pls.

SCOTESE, C. R. and MCKERROW, W. S. 1990. Revised world maps and introduction. 1–21. *In* MCKERROW, W. S. and SCOTESE, C. R. (eds). *Palaeozoic Palaeogeography and Biogeography*. Geological Society of London Memoir, **12**, 435 pp.

SENNIKOV, N. V., YOLKIN, E. A., PETRUNINA, Z. E., GLADKIKH, L. A., OBUT, N. G., IZOKH, N. G. and KIPRIYANOVA, T. P. 2008. *Ordovician–Silurian biostratigraphy and paleogeography of the Gorny Altai*. Publishing House SB RAS, Novisibirsk, 156 pp.

SHAW, F. C. 1968. Early Middle Ordovician Chazy trilobites of New York. *New York State Museum and Science Services Memoir*, **17**, 1–163.

—— 1974. Simpson Group (middle Ordovician) trilobites of Oklahoma. *Memoirs of the Paleontological Society*, no. 6, 54 pp., 12 pls. (supplement to *Journal of Paleontology*, **48** (5))

SHIRLEY, J. 1933. A redescription of the known British Silurian species of *Calymene* (s.l.). *Memoirs and Proceedings of the Manchester Literary and Philosophical Society*, **77**, 53–67, pl. 1.

SIVETER, D. J. 1985. The type species of *Calymene* (Trilobita) from the Silurian of Dudley, England. *Palaeontology*, **28**, 783–792.

—— 1996. Calymenid trilobites from the Wenlock Series (Silurian) of Britain. *Transactions of the Royal Society of Edinburgh, Earth Sciences*, **86**, 257–285.

SMITH, M. P. and RASMUSSEN, J. A. 2008. Cambrian–Silurian development of the Laurentian margin of the Iapetus Ocean in Greenland and related areas. *Geological Society of America Memoir*, **202**, 137–167.

—— —— HIGGINS, A. K. and LESLIE, A. G. 2004. Lower Palaeozoic stratigraphy of the East Greenland Caledonides. *Geological Survey of Denmark and Greenland, Bulletin*, **6**, 5–28.

ŠNAJDR, M. 1957. Klasifikace čeledě Illaenidae (Hawle a Corda) v českém starším paleozoiku. [Classification of the family Illaenidae (Hawle and Corda) in the Lower Palaeozoic of Bohemia]. *Sborník Ústředního Ústavu Geologického*, **23**, 125–254, pls 1–12. [English summary 270–284]

—— 1958. Několik nových rodů trilobite z čeledě Scutelluidae. *Věstník Ústředního Ústavu Geologického*, **33**, 177–184.

—— 1960. Studie o čeledi Scutelluidae (Trilobitae). *Rozpravy Ústředního Ústavu Geologického*, **26**, 1–265, 36 pls.

—— 1975. New Trilobita from the Llandovery at Hýstov in the Beroun area, central Bohemia. *Sborník Ústředního Ústavu Geologického*, **50**, 311–316.

—— 1976. New proetid trilobites from the Silurian and Devonian of Bohemia. *Věstník Ústředního Ústavu Geologického*, **51**, 51–55, 2 pls.

—— 1977. New genera of Proetidae (Trilobita) from the Barrandian, Bohemia. *Věstník Ústředního Ústavu Geologického*, **52**, 293–297.

—— 1980. Bohemian Silurian and Devonian Proetidae (Trilobita). *Rozpravy Ústředního Ústavu Geologického*, **45**, 1–324, 64 pls.

—— 1986. Bohemian representatives of the genus *Ceratocephala* Warder, 1838 (Trilobita). *Věstník Ústředního Ústavu Geologického*, **61**, 83–92.

SØNDERHOLM, M. and HARLAND, T. L. 1989. Franklinian reef belt, Silurian, North Greenland. 263–270. *In* GELDSETZER, H. H. J., JAMES, N. P. and TEBBUTT, G. E. (eds). *Reefs, Canada and adjacent areas*. Canadian Society of Petroleum Geologists Memoir, **13**, 775 pp.

—— —— DUE, P. H., JØRGENSEN, L. N. and PEEL, J. S. 1987. Lithostratigraphy and depositional history of Upper Ordovician–Silurian shelf carbonates in central and western North Greenland. *Rapports Grønlands Geologiske Undersøgelse*, **133**, 27–40.

STEININGER, J. 1831. *Bemerkungen über die Versteinerungen, welche in dem Übergangs Kalkgebirge der Eifel gefunden werden*. Eine Beilage zum Gymnasial-Programm zu Trier, Blattau, 46 pp.

STRUSZ, D. L. 1980. The Encrinuridae and related trilobite families, with a description of Silurian species from southestern Australia. *Palaeontographica, A*, **168**, 1–68.

SWINNERTON, H. H. 1915. Suggestion for a revised classification of trilobites. *Geological Magazine*, **2**, 487–496, 538–545.

TEICHERT, C. 1937. *Ordovician and Silurian faunas from Arctic Canada*. Report of the Fifth Thule Expedition 1921–24, The Danish expedition to Arctic North America in charge of Knud Rasmussen, Vol. I, no. 5. Gyldendalske Boghandel, Nordisk Forlag, Copenhagen, 169 pp., 24 pls.

THOMAS, A. T. 1978. British Wenlock trilobites. Part 1. *Palaeontographical Society Monograph*, **132**, 1–56, 14 pls.

—— 1979. Trilobite associations in the British Wenlock. 447–451. *In* HARRIS, A. L., HOLLAND, C. H. and LEAKE, B. E. (eds). *The Caledonides of the British Isles – Reviewed*. Geological Society of London, Special Publication **8**, xii + 768 pp.

—— 1981. British Wenlock trilobites. Part 2. *Palaeontographical Society Monograph*, **134**, 57–99, pls 15–25.

—— 1994. Silurian trilobites from the G. B. Schley Fjord region, eastern Peary Land, North Greenland. *Rapports Grønlands Geologiske Undersøgelse*, **164**, 29–35.

—— and HOLLOWAY, D. J. 1988. Classification and phylogeny of the trilobite order Lichida. *Philosophical Transactions of the Royal Society of London, Series B, Biological Sciences*, **321**, 179–262.

—— and LANE, P. D. 1999. Trilobite assemblages of the North Atlantic region. 444–457. *In* BOUCOT, A. J. and LAWSON, J. D. (eds). *Paleocommunities: a case study from the Silurian and Lower Devonian*. Cambridge University Press, Cambridge, 895 pp.

THOMAS, N. L. 1929. Hypoparia and opisthoparia from the St Clair Limestone, Arkansas. *Bulletin of the Scientific Laboratories of Denison University*, **24**, 1–25.

THORSLUND, P. 1930. Über einige neue Trilobiten aus dem älteren Letaena-Kalk Dalekarliens. *Bulletin of the Geological Institution of the University of Uppsala*, **22**, 299–303, pl .4.

—— 1940. On the Chasmops Series of Jemtland and Sodermanland (Tvären). *Sveriges Geologiska Undersökning, Afhandlingar och Uppsatser, Series C*, **436**, 1–191.

TÖRNQUIST, S. L. 1884. Undersökningar öfver Siljansområdets trilobitfauna. *Sveriges Geologiska Undersökning, Afhandlingar och Uppsatser, Series C*, **66**, 1–101.

TORSVIK, T. H., SMETHURST, M. A., MEERT, J. G., VAN DER VOO, R., MCKERROW, W. S., BRASIER, M. D., STUART, B. A. and WALDERHAUG, H. J. 1996. Continental break-up and collision in the Neoproterozoic and Palaeozoic – a tale of Baltica and Laurentia. *Earth Science Reviews*, **40**, 229–258.

TRIPP, R. P. 1954. Caradocian trilobites from mudstones at Craighead Quarry, near Girvan, Ayrshire. *Transactions of the Royal Society of Edinburgh*, **62**, 655–693, 4 pls.

—— 1962. Trilobites from the *confinis* Flags (Ordovician) of the Girvan district, Ayrshire. *Transactions of the Royal Society of Edinburgh*, **65**, 1–40, 4 pls.

—— 1967. Trilobites from the Upper Stinchar Limestone (Ordovician) of the Girvan district, Ayrshire. *Transactions of the Royal Society of Edinburgh*, **67**, 43–93, 6 pls.

—— 1980. Trilobites from the Ordovician Balchatchie and lower Ardwell groups of the Girvan district, Scotland. *Transactions of the Royal Society of Edinburgh*, **71**, 123–145.

VAN INGEN, G. 1901. The Siluric fauna near Batesville, Arkansas. II. Paleontology: Trilobitea. *Columbia University, School of Mines Quarterly*, New York, **23**, 34–74.

VOGDES, A. W. 1886. *Description of a new crustacean from the Clinton Group of Georgia, with remarks upon others*. Privately published, New York, 5 pp.

—— 1890. A bibliography of the Paleozoic crustacean from 1698 to 1889. *Bulletin of the US Geological Survey*, **63**, 1–177.

WARBURG, E. 1925. The trilobites of the Leptaena Limestone in Dalarne. *Bulletin of the Geological Institution of the University of Uppsala*, **17**, 1–446, 11 pls.

WARDER, J. A. 1838. New trilobites. *American Journal of Earth Science*, **34**, 377–380.

WEBER, V. N. 1932. Trilobites of the Turkestan. *Trudy Geologicheskogo Komiteta*, **178**, 3–151. [In Russian with English summary]

WELLER, S. 1907. The paleontology of the Niagaran limestone in the Chicago area. The Trilobita. *Bulletin of the Chicago Academy of Sciences*, **4**, 163–281, pls 16–25.

WESTROP, S. R. and EOFF, J. D. 2012. Late Cambrian (Furongian; Paibian, Steptoean) Agnostid Arthropods from the Cow Head Group, Western Newfoundland. *Journal of Paleontology*, **86**, 201–237.

—— and LUDVIGSEN, R. 1983. Systematics and paleoecology of the Upper Ordovician trilobites from the Selkirk Member of the Red River Formation, Southern Manitoba. *Manitoba, Department of Energy and Mines, Geological Report*, **GR 82–2**, 1–51.

—— 2000. The Late Cambrian (Marjuman) trilobite *Hysteropleura* Raymond from the Cow Head Group, western Newfoundland. *Journal of Paleontology*, **74**, 1020–1030.

—— and RUDKIN, D. M. 1999. Trilobite taphonomy of a Silurian reef: Attawapiskat Formation, Northern Ontario. *Palaios*, **14**, 389–397.

WHITEAVES, J. F. 1904. Preliminary list of fossils from the Silurian (Upper Silurian) rocks of the Ekwan River, and Sutton Mill Lakes, Keewatin, collected by D. B. Dowling in 1901, with description of such species as appear to be new. *Geological Survey of Canada, Annual Report*, **14**, 38F–59F.

WHITFIELD, R. P. 1878. Preliminary descriptions of new species of fossils from the lower geological formations of Wisconsin. *Annual Report of the Wisconsin Geological Survey* (for 1877), 50–89.

—— 1880. Descriptions of new species of fossils from the Paleozoic formations of Wisconsin. *Annual Report of the Wisconsin Geological Survey* (for 1879), 44–71.

—— 1882. Palaeontology. *Geology of Wisconsin, Wisconsin Geological Survey*, **4**, 161–363.

—— 1897. Descriptions of new species of Silurian fossils from near Fort Cassin and elsewhere on Lake Champlain. *Bulletin of the American Museum of Natural History*, **9**, 177–184.

WHITTARD, W. F. 1938. The Upper Valentian trilobite fauna of Shropshire. *Annals and Magazine of Natural History*, **11**, 85–140, 2 pls.

—— 1939. The Silurian Illaenids of the Oslo region. *Norsk Geologisk Tidsskrift*, **19**, 275–295, 4 pls.

—— 1961. The Ordovician trilobites of the Shelve Inlier, West Shropshire. *Monographs of the Palaeontographical Society*, **5**, 163–196, pls 22–25.

WHITTINGTON, H. B. 1950. A monograph of the British trilobites of the Family Harpidae. *Monographs of the Palaeontographical Society*. Publication No. 447, issued as part of Volume 103 for 1949, 55 pp., 7 pls.

—— 1954. Ordovician trilobites from Silliman's Fossil Mount. *Geological Society of America Memoir*, **62**, 119–149.

—— 1963. Middle Ordovician trilobites from Lower Head, western Newfoundland. *Bulletin of the Museum of Comparative Zoology, Harvard College*, **129**, 1–118, 36 pls.

—— 1965. Trilobites of the Ordovician Table Head Formation, western Newfoundland. *Bulletin of the Museum of Comparative Zoology, Harvard College*, **132**, 275–442, 68 pls.

—— 1997. Illaenidae (Trilobita): morphology of thorax, classification, and mode of life. *Journal of Paleontology*, **71**, 878–896.

—— 1999. Siluro-Devonian Scutelluinae (Trilobita) from the Czech Republic: morphology and classification. *Journal of Paleontology*, **73**, 414–430.

—— 2000. *Stygina, Eobronteus* (Ordovician Styginidae, Trilobita): morphology, classification, and affinities of Illaenidae. *Journal of Paleontology*, **74**, 879–889.

—— and EVITT, W. R. 1954. Silicified Middle Ordovician trilobites. *Memoirs of the Geological Society of America*, **59**, 1–137, 33 pls.

—— and KELLY, S. R. A. 1997. Morphological terms applied to Trilobita. 313–329. *In* KAESLER, R. L. (ed.). *Treatise on invertebrate paleontology. Part O. Arthropoda 1. Trilobita, revised. Volume 1: Introduction, Order Agnostida, Order Redlichiida.* Geological Society of America and University of Kansas Press, Boulder, Colorado and Lawrence, Kansas, xxiv + 530 pp.

WRIGHT, A. J. and CHATTERTON, B. D. E. 1988. Early Devonian trilobites from the Jesse Limestone, New South Wales, Australia. *Journal of Paleontology*, **62**, 93–103.

WU, H.-J. 1977. Comments on new genera and species of Silurian–Devonian trilobites in southwest China and their significance. *Acta Palaeontologica Sinica*, **16**, 95–119. [In Chinese with English summary]

XIA, S.-F. 1978. Ordovician trilobites. 157–185. *In* STRATIGRAPHICAL RESEARCH GROUP OF THE YANGTZE GORGE, GEOLOGICAL BUREAU OF HUBEI (ed.). *Sinian to Permian stratigraphy and paleontology of the East Yangtze Gorge area.* Geological Publishing House, Beijing, 381 pp. [In Chinese]

XIANG, L.-W. and ZHOU, T.-M. 1987. Trilobita. 294–335. *In* WANG, X.-F., XIANG, L.-W., NI, S. Z., ZENG, Q.- L., XU, G.-H., ZHOU, T.-M., LAI, C.-G. and LI, Z.-H. (eds). *Biostratigraphy of the Yangtze Gorges area, 2. Early Palaeozoic Era.* Geological Publishing House, Beijing, 641 pp.

YANISHEVSKY, M. 1918. On some representatives of the Upper Silurian fauna of the Caucausus. *Ezhegodnik Russkii Paleontologichestkogo Obshchestva*, **2**, 47–63, pl. 11. [In Russian with English summary]

YI, Y.-G. 1978. Silurian Trilobita. 265–269. *In* HUBEI PROVINCE BUREAU OF GEOLOGY, SANXIA STRATIGRAPHICAL RESEARCH GROUP (ed.). *Sinian to Permian Stratigraphy and Paleontology in the Xiadong Area.* Geological Press, Beijing, 381 pp. [In Chinese]

—— 1989. The Silurian stratigraphy and paleontology in Elangshan District, Sichuan. Description of selected fossils (4). Trilobita, *Bulletin of the Chengdu Institute of Geology and Mineral Resources*, **11**, 138–147. [In Chinese with English summary]

YIN, G.-Z. and LI, S.-J. 1978. Trilobita. 385–594. *In* GUIZHOU TEAM OF STRATIGRAPHY AND PALEONTOLOGY (ed.). *Paleontological Atlas of Southwest China, Guizhou Volume 1, Cambrian–Devonian.* Geological Publishing House, Beijing, 843 pp., 214 pls. [In Chinese]

ZHANG, W.-T. and SUN, X.-W. 2007. Silurian trilobites from Santanghu, Barkol, NE Xinjiang. *Acta Palaeontologica Sinica*, **46**, 33–44.

ZHOU, Z.-Y. and ZHANG, J.-L. 1978. Cambrian–Ordovician boundary of the Tangshan area with descriptions of the related trilobite fauna. *Acta Palaeontologica Sinica*, **17**, 1–28.

—— and ZHOU, Z.-Q. 1982. An Ashgill (Rawtheyan) trilobite faunule from Ejin Qi, Nei Monggol (Inner Mongolia). *Acta Palaeontologica Sinica*, **21**, 659–671. [In Chinese with English summary]

ZHOU, Z.-Q., SIVETER, D. J. and OWENS, R. M. 2000. Devonian proetid trilobites from Inner Mongolia, China. *Senckenbergiana Lethaea*, **79**, 459–499.